UNIVERSITY OF CALIFORNIA
BIBLIOGRAPHIC GUIDES

LITERATURE
OF
AGRICULTURAL
RESEARCH

by

J. RICHARD BLANCHARD

and

HARALD OSTVOLD

1958
UNIVERSITY OF CALIFORNIA PRESS
BERKELEY AND LOS ANGELES

University of California Press
Berkeley and Los Angeles
California
*
Cambridge University Press
London, England
*
© 1958 by The Regents of the
University of California
Library of Congress Catalog Card Number: 57-12942

PREFACE

The literature of agriculture and its related fields has grown tremendously in recent years. Agricultural experimentation and research have become major activities of almost every government and of thousands of commercial and nonprofit organizations. Hundreds of agencies, staffed by scientists and technicians, are pouring forth an increasing quantity of research results, issued in a confusing variety of forms and media. There are not only thousands of journals exclusively concerned with the literature of agriculture, but innumerable research reports and informative articles appear in serial publications and in bulletins not normally associated with agricultural topics. The problem of bibliographical control, of systematizing this superabundance of publications in such a way that facts can be obtained from then when called for, has become extremely complex.

This complexity arises from the inadequacy of control devices—bibliographies, indexes, and abstracting services—as well as from the quantity of publications. Such devices appear and disappear, overlap, skip years and whole subject areas, and omit the literature of entire countries and regions. It should be helpful, therefore, to take stock from time to time and to sum up descriptively these guides to information and literature. Such a summation may reveal weaknesses and gaps in the bibliographical machinery and thus provide a stimulus to improvement. It is hoped that the present work will be considered a step in this direction.

The literature has been arranged by broad subjects, the following headings having been selected as most likely to bring together related reference works: agriculture in general, plant sciences, animal sciences, physical sciences, food and nutrition, and social sciences. Within the subdivisions for sections, reference works have been arranged by the following categories: bibliographies of bibliographies and general works, abstracting journals, bibliographies and indexes, encyclopedias, dictionaries, directories, handbooks, yearbooks, history and biography, geography, abbreviations, periodical lists, societies and organizations, tables, and miscellaneous. Not all subjects have reference works in each of these categories, not is it always clear whether a certain work should be classified as an encyclopedia, a dictionary, or a handbook. Where general encyclopedic works seem to be lacking, some of the outstanding textbooks have been cited as useful sources of background information. Annotations have been provided when the title does not fully explain the nature of the work.

Some subjects commonly associated with the curriculum of agricultural colleges have not been included. The various fields of home economics, for instance, such as clothing, decorative arts, family living, and home planning, depend on the extensive literature of such widely separated subjects as sociology, fine arts, and psychology, and are not covered. Such fields as agricultural economics, rural sociology, and agricultural education, are covered only to the extent that it seemed practicable and possible to describe reference tools specifically written for them.

This is the first of a series of bibliographic guides sponsored by the University of California libraries and issued by the University of California Press. These guides, each of which will cover a specific subject, will be prepared in whole or in part by members of the library staffs. This work, which was started in the United States Dept. of Agriculture Library in 1948, represents more than eight years of sporadic labor by the two compilers. Practically every type of bibliographical and informational source has been used to find the references cited. Card catalogs and collections in the libraries of the Department of Agriculture, the University of Illinois, the University of California at Berkeley and at Davis, and the University of Minnesota were searched. Standard, current bibliographical tools, such as the <u>Bibliography of agriculture</u>, the <u>Agricultural index</u>, and <u>Biological abstracts</u>, were scanned to find useful works. In general, only the more valuable reference tools have been included, with the hope that they will guide the reader to further sources of information. Emphasis has naturally been placed on American publications, although an attempt has been made to list foreign works of first importance. Since directories are often issued irregularly, with many variations in title and method of publication, only the most recent volume examined has usually been cited. Exceptions to this rule have been made, however, in the few instances in which complete information about all issues of a directory was available. No attempt has been made to list all specialized bibliographies, statistical sources, and directories, simply because their numbers are overpowering and their use is often limited (see p. 8). More specific explanations of the selection policy will be found in the introductions to the different sections. Fortunately, it is generally recognized that a compilation of this type can rarely be complete. Suggestions and comments will, of course, be most gratefully received. Several recent titles, noted at the last minute, will be found in the Addendum, Section G.

Particular thanks are due Professor Ralph Shaw, formerly Director, U.S. Department of Agriculture Library; Mrs. Margaret Bryant, Head of that library's Bibliography Division; and Mr. Oliver Shipley of its Reference Section for much help and advice. Assistance has also been received from Mrs. Aileen Jaffa, Agricultural Reference Service, University of California, Berkeley; Miss Orpha Cummings, Librarian of the Giannini Foundation of the University of California; Mr. D. A. Brown, Librarian, College of Agriculture, University of Illinois; and Professor Emil Mrak of the University of California. Especial thanks are due Mrs. Carol Catlin for her careful work in preparing the manuscript.

The U.S. Department of Agriculture's <u>Bibliographic style, a manual for use in the division of bibliography of the library</u> (Bibliographical bulletin no. 16) has been used as a guide for bibliographical style. Words have been abbreviated according to the practice of the U.S. Department of Agriculture Library as set forth on pages 331-349 of its <u>List of serials currently received in the Library of the United States Department of Agriculture</u>. (Bibliographical bulletin no. 12.)

University of California, Davis   J.R.B.
University of Minnesota, St. Paul   H.O.

CONTENTS

SECTION A. AGRICULTURE: GENERAL 3

    Bibliographies of bibliographies, 3 - Abstracting journals, 5 - Bibliographies and indexes, 8 - Encyclopedias, 12 - Dictionaries, 13 - Directories, 15 - Handbooks and yearbooks, 17 - History and biography, 18 - Geography, 20 - Abbreviations, 21 - Periodical lists, 21 - Societies and organizations, 23 - Tables, weights, and measures, conversion factors, and statistical methods, 24 - Classification systems, 25

SECTION B. PLANT SCIENCES 27

  Botany 27

    Bibliographies of bibliographies, 27 - Abstracting journals and reviews, 28 - Bibliographies and indexes: early, 28 - Bibliographies and indexes: modern, 29 - Dictionaries, 33 - Directories, 35 - Flora: bibliographies, 35 - Flora: general, 36 - Economic botany, 36 - History and biography, 37 - Miscellaneous, 38

  Horticulture and Agronomy 39

    Abstracting journals and annual reviews, 40 - Bibliographies and indexes, 40 - Dictionaries, encyclopedias, and manuals, 42 - Directories and lists, 44 - Handbooks and texts, 45 - History, 46 - Periodical lists, 46

  Plant Breeding 46

    Abstracting journals, 46 - Bibliographies, 47 - Dictionaries, 48 - Directories, 48 - Handbooks and texts, 49

[vii]

Plant Pathology 49

    General works, 50 - Abstracting journals, 50 - Bibliographies and indexes, 51 - Geography, 53 - Handbooks, manuals, and texts, 53 - Fungi, 54

Forestry and Forest Products 55

    Abstracting journals, 55 - Bibliographies and indexes, 56 - Dictionaries, 58 - Directories, 60 - Trees: lists and descriptive guides, 62 - Handbooks and manuals, 63 - Geography, 64 - Periodical lists, 64 - Statistics, 64 - Miscellaneous, 65

SECTION C. ANIMAL SCIENCES 66

Economic Zoology 66

    Bibliographies of bibliographies, 66 - Abstracting journals, 67 - Bibliographies and indexes: general, 67 - Bibliographies and indexes: early, 68 - Bibliographies and indexes: special, 69 - Dictionaries, 70 - Nomenclatural guides, 71 - Directories, 71 - Geography, 72 - Periodical lists, 72 - Textbooks, 73

Animal Husbandry 73

    Abstracting journals, 73 - Bibliographies and indexes, 74 - Encyclopedic works, 75 - Dictionaries, 76 - Directories, 77 - Handbooks, texts, etc., 78 - History: general, 79 - History: cattle: individual breeds, 81 - History: sheep, 82 - History: swine, 82 - History: horses, 82 - Geography, 82 - Periodical lists, 83 - Miscellaneous, 83

Poultry Husbandry 83

    Abstracting journals, 84 - Bibliographies: general, 84 - Bibliographies: early, 85 - Bibliographies: special, 85 - Bibliographies: United States experiment station and extension service publications, 86 - Encyclopedias, 87 - Dictionaries, 87 - Breeds and species, 87 - Directories, 88 - Textbooks, 89 - Statistics, 89 - Miscellaneous, 90

Veterinary Medicine 90

    Bibliographies of bibliographies: general medicine, 90 - Abstracting journals, 91 - Bibliographies and indexes: general, 92 - Bibliographies and indexes: early, 92 - Bibliographies and indexes: library catalogs, 92 - Bibliographies and indexes: regional, 93 - Bibliographies and indexes: veterinary zoology and entomology, 93 - Bibliographies and indexes: special, 94 - Encyclopedias and handbooks, 94 - Dictionaries, 95 - Directories, biography, 96 - History, 96 - Periodical lists, 97

Economic Entomology 98

    Bibliographies of bibliographies, 98 - Abstracting journals, 99 - Bibliographies and indexes: general, 99 - Bibliographies and indexes: early, 101 - Bibliographies and indexes: library catalogs, 102 - Bibliographies and indexes: North American, 102 - Bibliographies and indexes: Canadian and United States government publications, 103 - Bibliographies and indexes: other countries, 104 - Bibliographies and indexes: insect control, 105 - Dictionaries, 105 - Nomenclatural guides, 106 - Directories, 106 - Handbooks and texts, 108 - History, 109 - Biography, 110 - Geography, 110 - Periodical lists, 110 - Miscellaneous, 111

Apiculture 111

    Bibliographies of bibliographies, 111 - Abstracting journals, 112 - Bibliographies, 112 - Encyclopedic works, 112 - Dictionaries, 113 - Directories, 113 - History, 114 - Periodical lists, 114

Pest Control 114

    Abstracting journals, 115 - Bibliographies and indexes, 115 - Dictionaries, 116 - Directories, 116 - Handbooks, manuals, texts, etc., 117 - Legislation, 118

Commercial Fishing and Fisheries 118

    Bibliographies of bibliographies, 118 - Abstracting journals, 119 - Bibliographies and indexes: general, 120 - Bibliographies and indexes: early, 120 - Bibliographies and indexes: special, 120 - Bibliographies and indexes: United States government publications, 121 - Bibliographies and indexes: angling, 121 - Bibliographies and indexes: translations, 122 - Bibliographies and indexes: motion pictures, 122 - Dictionaries and encyclopedias, 122 - Directories and yearbooks, 123 - Handbooks, texts, etc., 124 - History, 125 - Geography, 126 Periodical lists, 126 - Statistics, 126 - Miscellaneous, 127

SECTION D. PHYSICAL SCIENCES 128

    Agricultural Chemistry 128

        Abstracting journals and reviews, 128 - Bibliographies and indexes, 129 - Dictionaries and encyclopedias, 130 - Handbooks, manuals, etc., 131

    Soils and Fertilizers 131

        Abstracting journals, 132 - Bibliographies and indexes, 132 - Dictionaries, 134 - Directories, 135 - Handbooks and manuals, 135

Agricultural Engineering and Irrigation 136

    Abstracting journals, 136 - Bibliographies and indexes, 137 - Dictionaries, 138 - Directories, 138

Meteorology. 139

    Abstracting journals, 139 - Bibliographies and indexes, 140 - Glossaries, 140 - Handbooks and tables, 140

SECTION E. FOOD AND NUTRITION 142

    General works, reviews, and bibliographies of bibliographies, 142 - Abstracting journals, 143 - Bibliographies and indexes, 146 - Dictionaries and encyclopedias, 150 - Directories, 152 - Handbooks, manuals, etc., 156 - Periodical lists, 158 - Tables, 158

SECTION F. SOCIAL SCIENCES 160

Agricultural Economics, Statistics, and Legislation 160

    General works, 160 - Bibliographies and indexes, 161 - Statistics, 163 - Legislation, 165

Rural Sociology 166

    Abstracting journals, 166 - Bibliographies and indexes, 167 - Directories, 167 - Handbooks and texts, 167

Agricultural Education 167

    Bibliographies, 168 - History, 168 - Directories, 168 - Miscellaneous, 169

SECTION G. ADDENDUM 170

INDEX 175

LITERATURE OF AGRICULTURAL RESEARCH

SECTION A
AGRICULTURE: GENERAL

The reference works listed in this section are general compilations containing information relating to all branches of agriculture. Publications referring specifically to such subjects as soils and plant pathology are described in later sections. Since agriculture is an all-pervading and ubiquitous subject, it may also be necessary on occasion to use nonagricultural reference works such as the Encyclopedia Britannica and the Readers' guide to periodical literature. These and many other standard reference tools are carefully described in Constance M. Winchell's Guide to reference books (A8).

Bibliographies of Bibliographies

Although they do not attempt to describe all types of agricultural reference works, the publications noted below are helpful for the identification of separately published bibliographies. For the practicing agricultural librarian and research worker the specific compendiums of Frykholm (A3), Johnson (A6), and Schaeffer (A5) are of first importance. However, the comprehensive general work by Theodore Besterman (A1) and the Bibliographic index (A2) should not be overlooked. (See Section G, Addendum, for additional material.)

A1   BESTERMAN, T. A world bibliography of bibliographies and of bibliographical catalogues, calendars, abstracts, digests, indexes, and the like. Ed. 3. Geneva, Societies Bibliographica, 1955. 4v.
A monumental and uncritical list of separately published bibliographies. The author has strengthened his list of scientific and agricultural bibliographies in the third edition. Still incomplete for agriculture but nevertheless of the first importance.

A2   BIBLIOGRAPHIC index; a cumulative bibliography of bibliographies. 1937-   New York, Wilson, 1938-
Semiannual with annual and four-year cumulations. Lists alphabetically by subject separately published bibliographies and bibliographies issued in monographs and journals. Includes agricultural and scientific bibliographies in about 1,500 periodicals, including many in foreign languages.

A3   FRYKHOLM, L. Översikt över lantbruksforskningens och dess hjälpvetenskapers viktigare bibliografiska hjälpmedel. In Kungl.

Lantbruksakademiens Tidskrift 87(1948):459-489.
----- Supplement. In Kungl. Lantbruksakademiens Tidskrift 90(1951):257-271.

    A list of bibliographical reference works in agriculture prepared by the Librarian of the Royal Agricultural College Library in Uppsala, Sweden.

A4   INDEX bibliographicus; directory of current periodical abstracts and bibliographies, comp. by Theodore Besterman. Ed. 3. (UNESCO Publication no. 863. IFD Publication no. 247.) Paris, UNESCO; The Hague, International Federation for Documentation, 1952. 2 v.

    In French and English. V. 1, "Science and technology," is a useful guide to agricultural abstracting publications. However, certain publications listed have very little reference value, and a few items listed have ceased publication.

A5   INTERNATIONAL INSTITUTE of AGRICULTURE. Aperçu des bibliographies courantes concernant l'agriculture et les sciences connexes. A survey of current bibliographies on agriculture and allied subjects. [by Victor A. Schaeffer and others] Rome, 1937. 84 p.

----- Supplement. Nouvelles sources bibliographiques agricoles. New agricultural bibliographical sources. [by Sigmund von Frauendorfer] Rome, 1943. 44 p.

    Text in English and French. A compilation based on material received currently in the Library of the Institute. Each reference is accompanied by a note describing the contents of the bibliography. Arranged alphabetically by country with subject and title indexes. Translated into German as Die schrifttumsnachweise der landwirtschaftswissenschaft (Berlin, Parey, 1937), and Neue schriftumsnachweise aus dem gebiet der landwirtschaftswissenchaft und ihrer nebengebiete (Rome, 1943).

A6   JOHNSON, A. Bibliografiske hjelpemidler på landbrukets fagområder. Oslo, Cammermeyers Boghandel, 1942. 61 p.

    A selected list of agricultural reference works prepared by the Librarian of the Norwegian Agricultural College.

A6a   LAUCHE, R. Internationales handbuch der bibliographien des landbaus. World bibliography on agricultural bibliographies. Braunschweig, Forschungsanstalt fur Landwirtschaft. [In preparation, 1957.]

    When completed will be an important contribution to the bibliography of agriculture.

A7   REUNIÓN TÉCNICA de BIBLIOTECARIOS AGRÍCOLAS de AMÉRICA LATINA, TURRIALBA, 1953. Informe final, compilado por Carlos Víctor Penna, Relator General. Turrialba, Costa Rica, Instituto Interamericano de Ciencias Agrícolas. 2 v.

    Contains material presented at a meeting of Latin American librarians in 1953. The following articles and lists are of general reference value: "Coleccion basica - biblioteca agropecuaria" (a list of basic agricultural works for use in South American libraries), v. 1; "Bibliotecas agrícolas; una bibliografia tentativa" (largely concerns North and South American libraries), by Marietta Daniels and Angelina Martinez, v. 2, p. 57-72; "Lista de revistas agrícolas latinoamericanas preliminar," by Angelina Martinez and Noël James, v. 2, p. 92-115; and "Documentación bibliográfica en el campo de la agronomia y medicina veterinaria" (a brief survey of agricultural bibliography from an Argentine point of view), by Hans Gravenhorst, v. 2, p. 180-201.

A8 WINCHELL, C.M. Guide to reference books. Ed. 7. Chicago, American Library Association, 1951. 645 p.

----- Supplement, 1950-1952. Chicago, American Library Association, 1954. 117 p.

The most useful general guide in English to reference works. Includes chapters on science and applied science which have sections on agriculture, forestry, home economics, and botany.

Abstracting Journals

The abstracting journal is one of the chief tools in scientific research. Agriculture is well served by such journals, although several important services have been discontinued in recent years. The most serious loss was suffered when the Experiment station record (A22) suspended publication in 1946. As a result, the United States now has no abstracting journal devoted exclusively to agriculture, although Biological abstracts (A11) is supplying coverage for the plant and animal sciences. Abstracts for such important subjects as agricultural economics, statistics, and home economics are either not available or are issued in widely scattered journals. Other countries have also suffered. In Germany, for example, Der Forschungsdienst (A13), which contained an extensive agricultural abstract section, ceased publication in 1944. However, Dr. Oswald Asten's well-designed Agrar-Bibliographie (A9) is making a valiant effort to cover German literature. Dr. von Frauendorfer's Das Schriftum der Bodenkultur (A19) also provides a review of German and Austrian sources. The Landwirtschaftliches Zentralblatt (A15), started in 1955, should be of first importance. Perhaps the most useful agricultural abstracting services now being issued are those of the Commonwealth Agricultural Bureaux; these are described under the appropriate subjects in later sections of this work. (See Section G, Addendum, for additional material.)

A9 AGRAR-BIBLIOGRAPHIE; eine zusammenstellung der literatur auf dem gesamtgebiete der land- forst- und ernährungswirtschaft sowie der grund- und hilfswissenschaften. Bd. 1/2-   1945/46- Hamburg, Agrarwerbung G.M.B.H., 1948(?)-

Issued monthly. Devoted largely to the literature of German agriculture. Important items are abstracted. Classified arrangement. Annual author and subject indexes. A detailed statement concerning scope and purpose of this work is in Heft 12, 1954, p. 314-316.

A10 BERICHTE über die wissenschaftliche biologie. Referierendes organ der Deutschen Botanischen Gesellschaft. v. 1-   1926- Berlin, Springer, 1926-

Abt. A of Berichte über die gesamte Biologie. An international abstract journal for the biological sciences containing several sections of interest to agricultural science, such as: biochemistry, genetics, and applied botany. Comparable in several respects to Biological abstracts (A11).

Berichte über die gesamte Physiologie und experimentelle Pharmakologie, which is Abt. B of Berichte über die gesamte Biologie,

A. Agriculture: General

is also of interest to agriculturists since it contains sections on plant physiology, agricultural chemistry, food chemistry, colloid chemistry, etc.

A11 BIOLOGICAL abstracts; a comprehensive abstracting and indexing journal of the world's literature in theoretical and applied biology, exclusive of clinical medicine . . . Published (beginning with the literature of 1926) under the auspices of the Union of American Biological Societies. v. 1-   December 1926-   Philadelphia, 1926-

In its departments dealing with theoretical and applied bacteriology and botany this journal represents a continuation of <u>Abstracts of bacteriology</u> (B158) and <u>Botanical abstracts</u> (B5). Also includes material formerly abstracted in the <u>Experiment station record</u> (A22). Twelve issues per year plus annual author and subject indexes. Since January, 1954, monthly author indexes have been issued. Since 1939 it has been issued in sections, some of which were discontinued in 1953. The sections are as follows: Sec. A, Abstracts of general biology, 1939-   ; Sec. B, Abstracts of basic medical sciences (title varies), 1939-   ; Sec. C, Abstracts of microbiology, immunology, public health, and parasitology, 1939-   ; Sec. D, Abstracts of plant sciences, 1939-   ; Sec. E, Abstracts of animal sciences, 1939-   ; Sec. F., Abstracts of animal production and veterinary science, 1942-1953; Sec. G, Abstracts of food and nutrition research, 1943-1953; Sec. H, Abstracts of human biology, 1946-1953; Sec. J, Abstracts of cereal products, 1947-1953. Abstracts are prepared by specialists. An indispensable reference work for all workers in the biological sciences.

A12 DEUTSCHE landwirtschaftliche rundschau. v. 1-12, 1927-1935. Neudamm, 1927-1935.

A monthly general abstracting journal for agriculture. Continued by abstract section of <u>Der Forschungsdienst</u> (A13).

A13 DER FORSCHUNGSDIENST: Organ der deutschen landbauwissenschaft, v. 1-17, 1936-1944. Neudamm, 1936-1944.

Contains an extensive abstract section for general agricultural subjects. Continues <u>Deutsche landwirtschaftliche Rundschau</u> (A12).

A14 L'INSTITUT NATIONAL de la RECHERCHE AGRONOMIQUE. Series A. Annales agronomique. v. 1-   1950-   Paris, 1950-

The "Documentation" section abstracts selected world periodical literature concerning soils, fertilizers, plant physiology and pathology, and related subjects. The "Bibliographie" section reviews books and the "Sommaire des periodiques" section lists the contents of journals published throughout the world. Useful as a guide to French and European material. Series B, <u>Amélioration des plantes</u>, Series C, <u>Epiphyties</u>, and Series E, <u>Annales de technologie agricole</u> also have documentation sections which may be of interest to the plant scientist.

A15 LANDWIRTSCHAFTLICHES zentralblatt. 1955-   Berlin, Deutsche Akademie der Landwirtschaft, 1955-

A new agricultural abstracting journal, issued in 4 parts: Abt. 1, Landtechnik; Abt. 2, Pflanzliche produkte; Abt. 3, Tierzucht, tierernährung; Abt. 4, Veterinärmedizin. Is said to contain abstracts from some 1,000 periodicals prepared by about 750 scientists.

A16 NETHERLANDS. MINISTERIE van LANDBOUW, VISSERIJ en
WOEDSELVOORZIENING. Landbouwdocumentatie. Jaarg. 4-
1948- 's-Gravenhage, 1948-
A general weekly agricultural abstracting journal compiled by the
Centrum voor Landbouwdocumentatie, Wageningen. Arrangement
is according to the International decimal classification. Each issue
also includes the "Bulletin-catalogus" of the Bibliotheek der Landbouwhoogeschool, Wageningen.

A17 NORDISK jordbrugsforskning. Organ for nordiske jordbrugsforskeres
forening. [Scandinavian agricultural research. Organ for the
Association of Scandinavian agricultural research workers.]
Copenhagen and Stockholm, 1919-
The annual supplement, Förteckning över nordisk jordbrukslitter-
atur, is of reference value as a summary of Scandinavian agricultural literature. Has a classified arrangement with an author index.

A18 REVUE internationale des industries agricoles. v. 1- June 1939-
Paris, Commission Internationale des Industries Agricoles, 1939-
Abstracts world literature concerning agricultural industries, with
emphasis on food products. The lack of subject and author indexes
decreases the reference value. The annual "Table des matieres"
is too brief to be of much value.

A19 DAS SCHRIFTUM der bodenkultur. Bibliographische mitteilungen
aus dem gebiet der land- und forstwirtschaft und ihren
nebengebieten. Jahrg. 1- 1948- Vienna, Fromme, 1948-
Issued quarterly. A bibliographical review journal for agriculture,
forestry, and related subjects prepared in the Library of the
Hochschule für Bodenkultur in Vienna. Lengthy abstracts of important works. Articles in foreign and Austrian journals and new
books are listed according to subject.

A20 SPAIN. MINISTERIO de AGRICULTURA. Boletín bibliográfico agricola. no. 1- February 1948- Madrid, 1948-
Issued quarterly. Abstracts new books and lists journal articles
and bulletins in a classified arrangement. Covers world literature
with emphasis on material of interest to Spain. Apparently no indexes have been issued.

A21 TURRIALBA, Suplemento bibliografico. v. 4- January-March 1954-
Turrialba, Costa Rica, Instituto Interamericano de Ciencias
Agrícola, 1954-
The Suplemento bibliografico de Turrialba begins with v. 4, no. 1 of
Turrialba, which was started in 1950. It is published separately
and contains the following sections: (a) "Bibliographical index of
Latin American scientific agricultural journals." Indexes articles
published in Latin American agricultural journals received by the
Institute. (b) "Abstracts." The abstracts are given on 12.5 by
7.5 cm. cards. For certain subjects, such as cocoa, coffee, and
sugar, abstracts not only American publications but also those
published in other parts of the world. Photoprints may be request-
ed of the articles abstracted. (c) "List of acquisitions." Lists
new works received by the Library of the Institute.

A22 U.S. OFFICE of EXPERIMENT STATIONS. Experiment station
record. v. 1-95, September 1889-December 1946. Washington,
1890-1948. 95 v.
-----General index, v. 1-70, 1889-June 1934. Washington, 1903-1937. 6 v.

Primarily an abstract record of American agricultural experiment station and U.S. Department of Agriculture research and of articles by station and department men in nonofficial journals. Important foreign publications were also abstracted. Covers certain publications, such as the annual reports of the stations, which are not indexed in the Agricultural index (A23). Tables of contents for each volume list station and U.S. Department of Agriculture publications abstracted. Abstracts of important articles are detailed and were written by specialists. Original titles for foreign publications are given with English translation in brackets. Yearly indexes are by author and subject. General indexes, covering five or more years, by subject only. Still useful as a guide to agricultural literature published during the period covered.

Bibliographies and Indexes

The United States has two excellent indexing services, the Agricultural index (A23) and the Bibliography of agriculture (A24). General printed indexes such as these are described here. Specialized subject bibliographies are listed in their proper sections. It should be pointed out, however, that the number of specialized bibliographies—specialized as to locality, time, or subject—is so great that it is not possible to list them all. Virtually every book and article on agriculture includes a bibliography of some value, occasionally of unique value. The only publication that has attempted to systematize the approach to bibliographies appended to articles and books is the Bibliographic index (A2). Publications primarily intended to be bibliographical (which may be inferior to similar bibliographies appended to books or articles) are less numerous, but there are a great many of them. They may best be located by specific subject through the Bibliographic index (A2) or under appropriate headings in the annual subject index of the Bibliography of agriculture (A24). The subject indexes of the Experiment Station Record (A22) were particularly effective as locating devices for both bibliographical publications and bibliographies appended to books and articles. It can usually be assumed that an article or monograph on a specific subject will include a bibliography on the subject, which in turn will point to other articles with bibliographies, and so on. (See Section G, Addendum, for additional material.)

A23   AGRICULTURAL index; a cumulative subject index to agricultural periodicals, books, and bulletins. v. 1-   1916-   New York, Wilson, 1919-

      Issued monthly (except September). Monthly issues are superseded by cumulative volumes. Alphabetical subject index to more than a hundred periodicals and to bulletins, books, and publications of agricultural departments, experiment stations, extension services, agricultural societies, and other agencies. All publications now indexed are in English, most foreign publications having been dropped in 1949. "Some recent book titles," a section of each monthly issue, is useful for book selection. An indispensable reference work for any library serving agricultural interests.

A24   U.S. DEPARTMENT of AGRICULTURE. LIBRARY. Bibliography of agriculture. v. 1-   July 1942-   Washington, 1942-

Issued monthly. V. 1-2, issued in 6 sections: Agricultural economics and rural sociology, Agricultural engineering, Entomology, Plant science, Forestry, and Food processing and distribution. Beginning with v. 3, all sections were combined in a single publication and the scope of the bibliography broadened to cover the whole field of agriculture. Checklists of U.S. Department of Agriculture publications, state agricultural experiment station publications, and state agricultural extension service publications are provided. Classified by subject, indexed monthly by author and annually by author and subject. Subject and author indexes were provided semi-annually from 1943 to 1947. The most comprehensive single index to agricultural literature.

A25 ARGENTINE REPUBLIC. MINISTERIO de AGRICULTURA y GANDERIA de la NACION. DEPARTAMENTO de BIBLIOTECAS. Boletín bibliográfico. 1934-   Buenos Aires, 1934-
A classified list covering world agricultural literature.

A26 CLAY, E.W., comp. List of available publications of the United States Department of Agriculture. (U.S.D.A. Misc. P. 60, rev.) 1951. 173 p.
Lists by subject all publications of the department that are available either for free distribution or by purchase from the Superintendent of Documents. Gives information on distribution of publications and describes the various series and periodicals. Includes a list of depository libraries. Frequently revised.

A27 INTERNATIONAL INSTITUTE of AGRICULTURE. Bibliographie d'agriculture tropicale . . . Bibliography of tropical agriculture, 1931-1940. Rome, 1932-1942. 10 v.
Issued annually. Text in English and French. Covers primarily technical publications. References are annotated and classified by product. Indexed by author.

A28 INTERNATIONAL INSTITUTE of AGRICULTURE. Decennial index of publications (1930-1939). Rome, 1942. 55 p.
Pt. 1 is a classified list of all monographs, yearbooks, and periodicals published during this period. Pt. 2 indexes by author and title articles published in the English edition of the International review of agriculture. Pt. 3 is a general alphabetical index. Also issued in a French edition.

A29 INTERNATIONAL INSTITUTE of AGRICULTURE. Liste chronologique des publications éditées par l'Institut International d'Agriculture, non comprises dans la liste commerciale, et bibliographie de l'Institut depuis 1905. Rome, 1951. 19 p.
Pt. 1 lists statutes, conventions, periodicals, and other publications issued through 1939. Pt. 2 is a bibliography of official, semi-official, and private references on the origin and activities of the Institute.

A30 INTERNATIONAL INSTITUTE of AGRICULTURE. LIBRARY. Catalogue systématique: Classified catalogue. Rome, European Regional Office, 1948. 2747 p.
----- Suppléments 1-4. Rome, 1949-1950.
Pt. I, Catalogue des livres; Pt. II, Catalogue des bulletins; Pt. III, Catalogue des séries. Pt. I is arranged by the Universal decimal classification. Subjects are given in French. Has an alphabetical index of subjects and one of authors (including main entries).

A. Agriculture: General

A31   MCDONALD, D. Agricultural writers from Sir Walter of Henly to Arthur Young, 1200-1800. Reproductions in facsimile and extracts from their actual writings, enl. and rev. from articles which have appeared in "The Field" from 1903 to 1907. London, Cox, 1908. 228 p.
     Bibliography, chronologically arranged, of early British writings on agriculture. Publications are reviewed and quotations from them are frequently given. Brief biographical information about the authors is included. Illustrated with photographs of title pages, tables of contents, and pages of text.

A32   PERKINS, W.F., comp. British and Irish writers on agriculture. Ed. 2. Lymington, Eng., King, 1932. 193 p.
     List, alphabetically arranged, of books published from earliest times up to and including 1900. Gives for each: author, title, place of publication, date, and size. Lists of anonymous books, p.168-193.

A33   ROTHAMSTED EXPERIMENTAL STATION, Harpenden, Eng. LIBRARY. Library catalogue of printed books and pamphlets on agriculture published between 1471 and 1840. Ed. 2. Librarian, Mary S. Aslin. Aberdeen, Aberdeen University Press, 1940. 293 p.
     Has four sections: English authors and translations, Foreign authors and translations, Countries, and Incunabula. First two sections arranged alphabetically by author and by date; third arranged chronologically under country. Essential for identification of early agricultural books.

A34   SEL'SKOKHOZIAĬSTVENNAĬA literatura SSSR. [Agricultural literature of the USSR.] Moscow, 1926-
     Issued about six times annually. Prepared in the Tsentral'naia Nauchnaia Sel'skokhoziaĭstvennaia Biblioteka (Central Scientific Agricultural Library, Moscow): 1926-1928 issued under the title, Ezhegodnik agrarnoĭ literatury SSSR; 1929-1930 under the title, Agrarnaia literatura. References to Russian and foreign literature are arranged according to a decimal system set up by the International Institute of Documentation at Brussels. It is not known if this publication is still being issued. The U.S. Department of Agriculture Library received a 1951 issue in 1952 but none since.

A35   U.S. DEPARTMENT of AGRICULTURE. Bibliographical bulletin. no. 1- July 1943- Washington, 1943-
     These bibliographies each cover a special subject or topic; some are listed under appropriate sections elsewhere in this compilation.

A36   U.S. DEPARTMENT of AGRICULTURE. LIBRARY. Agricultural library notes. v. 1-17, no. 6, January 1926-June 1942. Washington, 1926-1942. 17 v.
     Continued by Bibliography of agriculture (A24). Lists principal library accessions, publications of U.S. Department of Agriculture workers issued outside the department, articles by state experiment station workers, translations, state extension publications, and bibliographies. Has an annual index.

A37   U.S. DEPARTMENT of AGRICULTURE. LIBRARY. Bibliographical contributions. no. 1-35, June 1919-September 1939. Washington, 1919-1939. 35 nos.
     The contributions cover various subjects. Some are listed in appropriate sections elsewhere in this compilation.

Agriculture: General 11

A38 U.S. DEPARTMENT of AGRICULTURE. LIBRARY. Library list.
no. 1-   June 1942-   Washington, 1942-
Of these lists, each dealing with a special subject, a few are noted under appropriate sections elsewhere in this compilation.

A39 U.S. DEPARTMENT of AGRICULTURE. LIBRARY. Selected list of American agricultural books and periodicals. (Libr. List 1, rev.) June 1954. 24 p.
Subject list of several hundred titles, with author index and list of publisher's addresses. Revised approximately every two years.

A40 U.S. DEPARTMENT of AGRICULTURE. OFFICE of INFORMATION. DIVISION of PUBLICATIONS. Index to the publications of the United States Department of Agriculture, 1901-1925, by M. A. Bradley, assisted by M. G. Hunt. Washington, 1932. 2689 p.

----- Supplements: 1926-1930, compiled by M. A. Bradley. 1935. 694 p.; 1931-1935, compiled by M. A. Bradley. 1937. 518 p.; 1936-1940, edited by M. A. Bradley. 1943. 763 p.
Author and subject index to all the printed publications of the department with the exception of the periodicals issued by the bureaus. The Journal of agricultural research and the Official record were included until they ceased publication. Indexed by minor topics as well as broad subjects.

A41 U.S. DEPARTMENT of AGRICULTURE. OFFICE of INFORMATION. DIVISION of PUBLICATIONS. List by titles of publications of the United States Department of Agriculture from 1840 to June 1901, inclusive; comp. and compared with the originals by R. B. Handy and M. A. Cannon. (U.S.D.A. B.6) 1902. 216 p.

----- Supplements: January 1901 to December 1925, inclusive, compiled by M. G. Hunt. (U.S.D.A. Misc. P. 9) 1927. 182 p.; January 1926 to December 1935, inclusive, compiled by M. G. Hunt. (U.S.D.A. Misc. P. 153) 1932. 46 p.; January 1931 to December 1935, inclusive, compiled by M. G. Hunt. (U.S.D.A. Misc. P. 252) 1936. 64 p.; January 1936 to December 1940, inclusive, compiled by M. H. Doyle. (U.S.D.A. Misc. P. 443) 1941. 68 p.; January 1941 to December 1945, inclusive, compiled by B. L. Zoeller and M. H. Doyle. (U.S.D.A. Misc. P. 611) 1946. 56 p.
Comprehensive lists of publications of the department since its inception arranged by number and date under the various bureaus and offices. No author or subject indexes. Lists covering publications issued from January 1936 to December 1945 are indexed by bureaus, series, periodical titles, etc.

A42 U.S. FOREIGN AGRICULTURAL SERVICE. Published information on agriculture in foreign countries, January 1937-December 1953. Washington, 1954. 65 p.
Classified index to information in periodicals, reports, and studies issued by the Foreign Agricultural Service, formerly the Office of Foreign Agricultural Relations. In three parts: 1, Regional information; 2, Commodity information; 3, Other studies.

A43 U.S. OFFICE of EXPERIMENT STATIONS. List of bulletins of the agricultural experiment stations in the United States from their establishment to the end of 1920. (U.S.D.A. Dept. B. 1199) 1924. 186 p.

----- Supplements: Calendar years 1921 and 1922, Sup. 1, by C. E. Pennington. (U.S.D.A. Dept. B 1199)1924. 24 p.; calendar years 1923

12   A. Agriculture: General

and 1924, Sup. 2, by C. E. Pennington. (U.S.D.A. Dept. B. 1199) 1926. 54 p.; calendar years 1925 and 1926, Sup. 3, by C. E. Pennington. (U.S.D.A. Dept. B. 1199) 1926. 62 p.; calendar years 1927 and 1928, by C. E. Pennington. (U.S.D.A. Misc. P. 65) 1930. 78 p.; calendar years 1929 and 1930, by C. E. Pennington. (U.S.D.A. Misc. P. 128) 1932. 88 p.; calendar years 1931 and 1932, by C. E. Pennington. (U.S.D.A. Misc. P. 181) 1934. 77 p.; calendar years 1939 and 1940, by C. E. Pennington. (U.S.D.A. Misc. P. 459) 1941. 86 p.; calendar years 1941 and 1942, compiled by H. V. Barnes. (U.S.D.A. Bibliog. B. 4) 1944. 70 p.
The lists are confined primarily to the regular bulletin series of the state experiment stations, including those of Alaska and the insular possessions. Circulars, annual reports, except those numbered as bulletins, and ephemeral publications are omitted. Arrangement is by state and numbers. Author and subject indexes are included.

A44   U.S. SUPERINTENDENT of DOCUMENTS. List of publications of the Agriculture Department, 1862-1902, with analytical index. (Bibliog. of U.S. Pub. Documents. Dept. List no. 1.) 1904. 623 p.
Arranged by date or number under the several bureaus, offices, and divisions of the department. Classification numbers used by the Superintendent of Documents are included. An appendix (p. 615-622) lists reports which originated in the Agriculture Department but which were printed only as Congressional documents.

A45   WAGENINGEN, NETHERLANDS. BIBLIOTHEEK der LANDBOU-WHOOGESCHOOL. Bulletin-catalogus. [Bulletin-catalog of the Library of the Agricultural College at Wageningen, Netherlands.] Wageningen, 1919-
A classified library accession list. Early issues each covered a particular field of agriculture but recent issues note only current accessions.

A46   ZIMMERMAN, F. L. and READ, P. R., comps. Numerical list of current publications of the United States Department of Agriculture, comp. by comparison with the originals. (U.S.D.A. Misc. P. 450) 1941. 939 p.
Compiled for the use of institutions and persons engaged in informational work. Strictly numerical arrangement without regard to series. Useful in identifying publications when only the number is known.

Encyclopedias

There are few modern agricultural encyclopedias worthy of note, although several good ones were published in the first part of the century. This can, no doubt, be attributed to the advance of science, which rapidly antiquates any general compilation. (See Section G, Addendum, for additional material.)

A47   BAILEY, L. H. Cyclopedia of American agriculture; a popular survey of agricultural conditions, practices, and ideals in the United States and Canada. New York, Macmillan, 1907-1909. 4 v.
Arranged by broad subjects with index for each volume. Signed articles by specialists, with bibliographies. Now too antiquated for

general purposes but useful historically. In 1917 a reprint with no change in text was issued as a "new edition." V. 2, "Crops" and v. 3, "Animals" were reissued in 1922 as separate works entitled Cyclopedia of farm crops and Cyclopedia of farm animals; there was no change in text, however.

A48   HUNTER, H., ed. Baillière's encyclopaedia of scientific agriculture. London, Baillière, Tindall & Cox, 1931. 2 v.
Arranged alphabetically with cross references linking various relevant subjects. Principal articles contributed by recognized authorities. Deals largely with British agriculture; but is applicable to Canada, New Zealand, Australia, and sections of the United States.

A49   NOUVEAU Larousse agricule, publié sous la direction de Raymond Branconnier ... et de Jacques Glandard. Paris, Librairie Larousse, 1952. 1,152, 78, XVI p.
A modern illustrated encyclopedia with a classified arrangement and a dictionary index.

A50   PAREY'S landwirtschafts-lexikon. Ed. 7. Hrsg. von L. W. Ries. Hamburg and Berlin, Parey, 1956. 2 v.
Well-illustrated encyclopedia covering the whole field of agriculture. Sixth ed., by W. Borgman, E. U. Brödermann, and P. Gisevius, has title: Illustriertes landwirtschafts Lexikon.

A51   WILCOX, E. V. Modern farmers' encyclopedia of agriculture; a compendium of farm science and practice ... New York, Judd, 1952. 543 p.
Presents in one volume summaries of scientific investigation in experiment stations, state and federal departments of agriculture, etc. Material is arranged in 8 sections: Field crops; Garden crops; Fruits and nuts; Beef cattle and dairying; Other livestock; Poultry; Drainage, fertilization, irrigation, soils; Miscellaneous. Indexed by subject.

A52   WRIGHT, R. P., ed. Standard cyclopedia of modern agriculture and rural economy. London, Gresham [1914?] 12 v.
An old but excellent compilation still useful for historical information. Principal articles by specialists. No bibliographies.

Dictionaries

Dictionaries are important and necessary: each branch of agriculture has its own language which must be explained to the layman and to workers in other fields. The most useful are bilingual or polylingual; these assist greatly in translation, which is essential to maintain the international flow of information. The fact that scientists are definitely inhibited by foreign languages is illustrated by the bibliographical citations in two similar journals, Soil science and Landwirtschaftliche Jahrbücher, for 1939. In the American journal 80.3 per cent of the references were in English, in the German periodical 92 per cent were in German. It is obvious that widespread development of translating services or world-wide adoption of a universal language would be of great benefit to research, but until such a millennium, the interlingual dictionary must be used. A list of such works is the first item noted on the following page. (See Section G, Addendum, for additional material.)

A. Agriculture: General

A53 UNITED NATIONS EDUCATIONAL, SCIENTIFIC and CULTURAL ORGANIZATION. Bibliography of interlingual scientific and technical dictionaries. Ed. 3. Paris, 1953. 178 p.
   A classified list including some, but not all, of the important interlingual dictionaries in agriculture and related fields. Has subject, author, and language indexes.

A54 BADING, H. Wörterbuch der landwirtschaft. Hamburg, Park Verlag, 1947. 156 p.
   Title page and preface in German and English. German-English, English-German dictionary of technical terms in agriculture and the more important food industries, designed for the use of farmers, gardeners, breeders, seed-growers, beekeepers, veterinary surgeons, agricultural colleges, etc. Tables of dimensions and weights, p. 154-156.

A55 BEZEMER, T. J. Dictionary of terms relating to agriculture, horticulture, forestry, cattle breeding, dairy industry, and apiculture, in English, French, German, and Dutch. Baltimore, Williams & Wilkins, 1935. 4 parts in 1 v.
   Multilingual dictionary devoted exclusively to agriculture. In 4 parts: one for each language. Terms in each language are given with equivalents in the other three, but are not defined. An edition with title page, preface, and foreword in Dutch was published in the Netherlands in 1934.

A56 CHAPMAN, D. H. A farm dictionary. London, National Federation of Young Farmers' Clubs, 1953. 209 p.
   An interesting compilation of British farm terms. Has some amusing definitions in the Johnsonian manner.

A57 DEVRIES, L. French-English science dictionary for students in agricultural, biological, and physical sciences, with a supplement of terms in aeronautics, electronics, radar, radio, television. Ed. 2. New York, McGraw-Hill, 1951. 596 p.

A58 DEVRIES, L. German-English science dictionary for students in chemistry, physics, biology, agriculture, and related sciences. Ed. 2, rev. and enl. New York, McGraw-Hill, 1946. 558 p.
   Revision and enlargement of an earlier edition, published in 1939. The number of terms, idioms, and abbreviations has been substantially increased.

A59 DZIEDUSZYCKI, A. Agricultural dictionary, English-Polish and Polish-English. For farmers, livestock breeders, tractor drivers, dairymen, gardeners, beekeepers, and foresters. Rev. and augmented by J. Andrew. London, F. P. Agency, 1945. 84 p.
   One alphabet, with English and Polish in parallel columns. Terms included are not defined. Title page in Polish and English.

A60 GEMINOVA, N. F., KRASNOSSELSKAJA, T. C., and USSOWSKY, B. N. English-Russian agricultural dictionary, ed. by T. A Krasnosselskaja. Moscow, Gostechizdat, 1944. 372 p.
   Contains 20,000 terms on plant industry, fruit culture, mechanization of agriculture, botany, beekeeping, etc. Illustrations of vegetables, tools, and farm machinery are given on p. 333-372. Title page in English and Russian.

A61 WOLFE, L. S. Farm glossary; definitions of words and phrases usually encountered in reading literature dealing with agriculture and its allied sciences. Orangeburg, S. C., 1948. 360 p.

Agriculture: General 15

Compilation of words and phrases used in farm papers, state and federal bulletins, and text books on agriculture. Biographical accounts of persons who have contributed to sciences closely related to agriculture are included as are those of mythological personages having direct connection with the lore of rural life.

Directories

A few general directories are noted below. Unfortunately, many agricultural directories of international scope are now badly out-of-date and need to be revised. Because it is almost impossible to obtain exact bibliographic information about directories that are issued periodically—many which are purportedly annuals are issued only at irregular intervals, some are issued regularly but with varying titles, some may be issued as separates in one year and as supplements to journals in the next—only the most recent volume that it has been possible to examine has been listed, except for those publications which have followed a consistent policy and for which this information was readily available.

A62 ANNUAIRE national de l'agriculture, 1945-1946. Ed. 8. Paris, Horizons de France, 1946. 595 p.
   French agricultural directory showing the organization of the Ministry of Agriculture and giving names and addresses of officers and employees. Lists of schools, institutions, associations, and dealers in farm supplies and equipment are included.

A63 COMMONWEALTH AGRICULTURAL BUREAUX. EXECUTIVE COUNCIL. Gazetteer of agricultural and forestry research stations in the British Commonwealth, 1952. Farnham Royal, Eng., 1952. 517 p.
   Supplementary to A64. Also includes research stations in the Republics of Ireland and India. There are alphabetical indexes to research stations and to subjects of research.

A64 COMMONWEALTH AGRICULTURAL BUREAUX. EXECUTIVE COUNCIL. List of research workers, 1956. Research workers in agriculture, animal health, and forestry, in the British Commonwealth, the Republic of the Sudan and the Republic of Ireland. Farnham Royal, Eng., 1956. 472 p.
   Lists names and addresses of all workers in the countries, colonies, protectorates, and mandated territories. Arranged by countries, colonies, etc. Includes an alphabetical index of names and of institutions.

A65 CONFÉDÉRATION EUROPÉENE de l'AGRICULTURE. Vade-mecum des principales organisations internationales tenant compte spécialement de leurs rapports avec l'agriculture et des organisations agricoles internationales et institutions internationales apparentées à l'agriculture. Aide-mémoire rédigé par Andre L. Geisendorf. (Publications de la CEA. Fasc. 5.) Brugg, Switzerland, 1951. 125 p.
   A directory of international agricultural organizations. Information concerning the history, purpose, structure, and membership of each group is provided.

A66 COUNTY agents' directory. 1915-   Chicago, Agricultural

A. Agriculture: General

Leaders' Digest, 1915-
Issued annually. Primarily a directory, arranged alphabetically by states, of county agricultural and home demonstration agents. Contains also a list of national organizations, officers of the American Farm Bureau Federation, and members of the National Live Stock Producers' Association. Various tables, showing sowing rates for field seeds and grasses, animal gestation periods, etc., are included. (Ed. 41, 1956. 224 p.)

A67 DAVIS, M. V. Guide to American business directories. Washington, Public Affairs Press, 1948. 242 p.
Classified list giving frequency of issue, name and address of publisher, and price. Each entry is accompanied by a brief descriptive contents note. Food industry and trade directories are noted on p. 96-126.

A68 DIRECTORY of biological laboratories. Ed. 5, 1951. Chicago, Burns Compiling and Research Organization, 1951. 164 p.
Lists laboratories in the United States and Canada. Largely confined to information about laboratories working in food and nutrition, pharmaceuticals, and organic chemicals. Clinical laboratories and those connected with schools, colleges, or universities are not included.

A69 INTERNATIONAL INSTITUTE of AGRICULTURE. Bibliothèques agricoles dans le monde et bibliothèques spécialisées dans les sujets se rapportant à l'agriculture. International directory of agricultural libraries and of libraries specialized in subjects related to agriculture. Rome, 1939. 311 p.
French and English. Arranged by continents and alphabetically by countries. For each library, subject specialities, date founded, size of collection, number of staff members, etc., are given. Geographical index is included.

A70 INTERNATIONAL INSTITUTE of AGRICULTURE. Les institutions d'expérimentation agricole dans les pays tempérés. Rome, 1933. 306 p.
Arranged by continents and alphabetically by countries. Includes a brief account of the organization of research in each country and gives information such as date of founding, location, size of staff, scope of work, etc., for each station. Has indexes of names of technicians and cities mentioned in the text.

A71 INTERNATIONAL INSTITUTE of AGRICULTURE. Les institutions d'expérimentation agricole dans les pays chauds. International directory of agricultural institutions in hot countries. Ed. 2. Rome, 1934. 563 p.
French and English. Arranged by continents and alphabetically by countries. Information given for each station includes geographical position, date of founding, size of staff, scope of work, work completed and in progress, and language used in correspondence. Name and station indexes are included.

A72 NATIONAL RESEARCH COUNCIL. Industrial research laboratories of the United States, comp. by James F. Maulk. Ed. 10. (Natl. Res. Council P. 379) 1956. 560 p.
Lists 4,834 laboratories. The index to research activities lists many subjects of interest to agriculture and related fields.

Agriculture: General 17

A73 UNITED NATIONS EDUCATIONAL, SCIENTIFIC and CULTURAL ORGANIZATION. Directory of institutions engaged in arid zone research. Paris, 1955. 110 p.
A partial list of institutions throughout the world engaged in arid zone research, such as irrigation, water resources, geology, plant breeding, etc.

A74 UNITED NATIONS EDUCATIONAL, SCIENTIFIC and CULTURAL ORGANIZATION. Directory of international scientific organizations. Paris, 1950. 217 p.
Includes international agricultural organizations.

A75 U.S. AGRICULTURAL MARKETING SERVICE. Directory of state departments of agriculture, 1955. Washington, 1955. 55 p.
Lists state marketing service agencies and the names of key officials of the state departments of agriculture. Also includes information concerning departments of agriculture in Alaska, Hawaii, and Puerto Rico and officials of the major associations of agricultural workers.

A76 U.S. DEPARTMENT of AGRICULTURE. Directory of organization and field activities of the Department of Agriculture: 1955. (Agr. Handb. 76) 1954. 198 p.
Organization section is arranged by bureau. Field activities are listed geographically. The 1950 edition (Agr. Handb. no. 12) was the last to list the names of field personnel.

A77 U.S. DEPARTMENT of AGRICULTURE. Officials of the U.S. Department of Agriculture. Washington, 1954. 2 p.
Lists only top officials. Revised regularly.

A78 U.S. DEPARTMENT of AGRICULTURE. Workers in subjects pertaining to agriculture in land-grant colleges and experiment stations, 1954-1955. (Agr. Handb. 78) 1955. 211 p.
Issued annually. Arranged alphabetically by states. Has an index of names.

A79 U.S. OFFICE of EXPERIMENT STATIONS. Agricultural research institutions and library centers in foreign countries, comp. by H. L. Knight. Rev. July 1, 1934. Washington, 1934. 64 p.
List of about 1,400 institutions, of which about 400 are classified as "primary research centers." Arranged alphabetically by countries. Gives only name and location of each. Includes an index of countries.

Handbooks and Yearbooks

A80 FARMING and mechanised agriculture. Ed. 4. Sir R. George Stapledon, Advisory Editor. London, Todd Reference Books, 1950. 407 p.
A general reference work and manual of information for British agriculture. Has reports on recent developments in agriculture; information about governmental and private organizations, institutions, and committees concerned with agriculture; statistics and tables; and a selected list of agricultural books, periodicals, and films. Index.

A81 U.S. DEPARTMENT of AGRICULTURE. Yearbook of agriculture, 1894- Washington, 1895-
Issued annually. Title varies. Volumes published before 1936

contain statistical information and summaries of developments in agriculture during the year. Since 1936 each volume has been devoted to a special subject, as follows: 1936, Better plants and animals, I; 1937, Better plants and animals, II; 1938, Soils and men; 1939, Food and life; 1940, Farmers in a changing world; 1941, Climate and man; 1942, Keeping livestock healthy; 1943-47, Science in farming; 1948, Grass; 1949, Trees; 1950-1951, Crops in peace and war; 1952, Insects; 1953, Plant diseases; 1954, Marketing; 1955, Water; 1956, Animal diseases. Statistics are now published annually in <u>Agricultural statistics</u> (F19). Indexes: 1894-1900, 1901-1905, 1906-1910, 1911-1915.

A82 WHITE, J.M. The farmer's handbook. Norman, University of Oklahoma Press, 1948. 440 p.
A handbook of ready reference on crops, farm animals, farm engineering, and other major phases of farming.

History and Biography

A83 AMERICAN men of science; a biographical directory. Ed. 9. New York and Lancaster, Pa., R. R. Bowker and The Science Press, 1955-1956. 3 v.
The majority of prominent American scientists working in agriculture and related fields are listed in v. 2, Biological sciences. V. 1, Physical aciences; v. 3, Social and behavorial sciences. First edition published in 1906.

A84 BIDWELL, P.W. and FALCONER, J.I. History of agriculture in the northern United States, 1620-1860. (Carnegie Institution of Washington, Pub. no. 358) Washington, Carnegie Institution of Washington, 1925. 512 p.
Fifth in a series of contributions to American economic history (See also A88). Subject treated in four parts: Pt. 1, Agriculture in the earliest settlements; Pt. 2, Rural economy in the eighteenth century; Pt. 3, Expansion and progress, 1800-1840; Pt. 4, Period of transformation, 1840-1860. Contains statistical tables on prices, exports, etc., of farm products and a "classified and critical bibliography," as well as numerous maps and charts. Indexed.

A85 EDWARDS, E.E. and RASMUSSEN, W.D., comps. A bibliography on the agriculture of the American Indians. (U.S.D.A. Misc. P. 447) 1942. 107 p.
A classified and annotated bibliography of 841 references with detailed index.

A86 EDWARDS, E. E. Bibliography of the history of agriculture in the United States. (U.S.D.A. Misc. P. 84) 1930. 307 p.
Comprehensive annotated bibliography classified by regions and products, states, and special subjects such as transportation and markets, farm implements and machinery, and agencies promoting agriculture, with an index. The section on agricultural leaders is useful for biography. "Only in the sections pertaining to agriculture in the strict sense of the word has an effort been made to list more than the main references. A few publications issued prior to 1900 and after 1929 are given, but only the literature printed during the intervening years has been carefully searched."

A87   GRAS, N. S. B. A history of agriculture in Europe and America. Ed.
        2. New York, Crofts, 1940. 496 p.
A88   GRAY, L. C. History of agriculture in the southern United States to
        1860, by Lewis Cecil Gray, assisted by Esther Katherine
        Thompson, with an introductory note by Henry Charles Taylor.
        (Carnegie Institution of Washington, Pub. no. 430.) Washington,
        Carnegie Institution of Washington, 1933. 2 v.
    Seventh in a series of contributions to the economic history of the
    United States. Material organized in 7 parts: Pt. 1, Agricultural
    beginnings and geographic expansion; Pt. 2, Agricultural industries
    in the colonial period; Pt. 3, Institutional development in the co-
    lonial period; Pt. 4, Economic evolution in the South; Pt. 5, The
    development of national economy; Pt. 6, Agricultural industries and
    husbandry in the post colonial period; and Pt. 7, Geographic expan-
    sion and regional development. Has a comprehensive bibliography
    and footnotes. Contains statistical tables on population and produc-
    tion, exports, and prices of typical southern commodities to 1860,
    and maps and charts. Indexed.
A89   HARDING, T. S. Two blades of grass, a history of scientific develop-
        ment in the U.S. Department of Agriculture. Norman, Univer-
        sity of Oklahoma Press, 1947. 352 p.
A90   HISTORIA agriculturae. Yearbook issued by Het Nederlands
        Agronomisch-Historisch Instituut. Deel 1-   1953-
        Groningen, Wolters, 1953-
    Each issue contains a world bibliography of books and articles on
    agricultural history published during the previous year. Edited by
    J. M. G. Van Der Poel.
A91   R U S. A biographical register of rural leadership in the United States
        and Canada, 1918-1930. Ithaca, N. Y., 1918-1930. 4 v.
    Still useful, as there is no current biographical compilation ex-
    clusively for agriculture. Last issue contains 6,881 names of
    prominent farmers, teachers, investigators, county and home dem-
    onstration agents, etc. Biographical information on many contem-
    porary leaders in agriculture can be found in Who's who in America,
    American men of science (A83), and other compilations.
A92   SCHMIDT, L. B. Topical studies and references on the economic
        history of American agriculture. Rev. ed. Philadelphia,
        McKinley, 1923. 126 p.
    Guide to the study of agricultural history. The bibliographical sec-
    tion, containing references to books, articles, and original sources,
    is classified and covers the period 1492-1923.
A93   TRUE, A. C. A history of agricultural experimentation and research
        in the United States, 1607-1927, including a history of the United
        States Department of Agriculture. (U.S.D.A. Misc. P. 251) 1937.
        321 p.
    A general survey, describing the organization of research and ex-
    perimentation and its general character. Gives the history of the
    Hatch Experiment Station Act of 1887 and various state experiment
    stations. Has a bibliography of 327 references. Author and subject
    indexes are included.
A94   U.S. BUREAU of AGRICULTURAL ECONOMICS. A chronology of
        American agriculture, 1790-1940. Washington, 1941. 1 sheet.
    A graphic chart.

A95 U.S. CONGRESS. HOUSE. COMMITTEE on AGRICULTURE. Research and related services in the United States Department of Agriculture. Washington, 1951. 4 v.
> A comprehensive survey of research and related activities carried on within the Department of Agriculture and by the department in coöperation with state land-grant colleges and other state agencies. For each subject of research, information is given concerning current work, history of the work, and additional work needed.

A96 U.S. DEPARTMENT of AGRICULTURE. Some landmarks in the history of the Department of Agriculture. (Agr. Hist. Ser. 2, rev.) 1951. 113 p.
> References consulted, p. 111-113.

A97 U.S. DEPARTMENT of AGRICULTURE. OFFICE of INFORMATION. Biographies of persons in charge of federal agricultural work, 1836 to date. (Document 3) 1953. 17 p.
> Has biographical sketches of all United States secretaries of agriculture or their equivalents.

A98 U.S. DEPARTMENT of AGRICULTURE. OFFICE of INFORMATION. Condensed history of the U.S. Department of Agriculture. (Document 4, rev.) 1950. 15 p.

A99 U.S. DEPARTMENT of AGRICULTURE. OFFICE of INFORMATION. Origin, structure, and functions of the U.S. Department of Agriculture. (Document 1) 1953. 26 p.

Geography

A100 BACON, L. B., and others. Agricultural geography of Europe and the Near East. (U.S.D.A. Misc. P. 665) 1948. 67 p.
> Series of maps (some colored), with accompanying text showing the geographic origin of food and other important agricultural products together with information on the climate, vegetation, and soil conditions in Europe.

A101 GERMANY. STATISTISCHES REICHSAMT. Deutscher landwirtschaftsatlas. Berlin, Hobbing, 1934. 32 p., 104 maps.
> Text and series of maps showing cultivated area of crops, numbers of livestock, private forests, etc.

A102 HAINSWORTH, R. G. A graphic summary of world agriculture. (U.S.D.A. Misc. P. 705) 1949. 51 p.
> Contains maps illustrating geographic distribution of agricultural production and trade, world population, and such physical factors as land forms, vegetation, soils, and annual rainfall.

A103 KLAGES, K. H. W. Ecological crop geography. New York, Macmillan, 1942. 615 p.
> A widely used text. Selected references at the end of each chapter.

A104 THORNTHWAITE, C. W. Atlas of climatic types in the United States, 1900-1939. (U.S.D.A. Misc. P. 421) 1941. 7 p., 96 plates.
> U.S. Soil Conservation Service and U.S. Weather Bureau coöperated to produce this series of maps, limited to a presentation of the relation of moisture to climate, and showing the normal position and variations from year to year of the principal climatic types. Literature cited and a list of 123 selected references are included.

A105 U.S. DEPARTMENT of AGRICULTURE. Atlas of American agriculture. Physical basis including land relief, climate, soils, and natural vegetation of the United States. Prep. under the supervision of O. E. Baker. Washington, 1936. Various paging.
Originally issued in parts: Land relief by F. J. Marschner. 1936. 6 p.; Climate—temperature, sunshine, and wind by J. B. Kincer. 1928. 34 p.; Climate—frost and the growing season by W. G. Reed, Climate—precipitation and humidity by J. B. Kincer. 1922. 48 p.; Soils of the United States by C. F. Marbut. 1935. 98 p.; Natural vegetation—grassland, desert shrub, and forests by H. L. Shantz and R. Zon. 1924. 29 p.
Collection of maps with accompanying text. Includes illustrations and bibliographies.

A106 U.S. DEPARTMENT of AGRICULTURE. Geography of the world's agriculture, by V. C. Finch and O. E. Baker. Washington, 1917. 149 p.
Series of maps, with text, showing acreage and production of principal crops and distribution of the livestock industry. List of statistical references, p. 148-149.

A107 VAN ROYEN, W. Atlas of the world's resources. V. 1: The agricultural resources of the world. New York, Prentice Hall, 1954. 254 p.
A most important work prepared in coöperation with the Bureau of Agricultural Economics. Contains basic information and statistics concerning world crops and livestock illustrated with carefully drawn maps. An important feature is the list of references for each section and the general list of selected references which includes a list, "Atlases containing maps pertaining to agricultural resources," on p. 257. V. 2 of the Atlas deals with mineral resources and v. 3, which is in preparation, will cover forest and fishery resources.

Abbreviations

A108 WHITLOCK, C., comp. Abbreviations used in the Department of Agriculture for titles of publications. (U.S.D.A. Misc. P. 337) 1939. 278 p.
Supersedes Abbreviations employed in Experiment station record for titles of publications (Dept. B. 1330). Entries arranged alphabetically, with abbreviations and titles being given in opposite columns. List of abbreviations of single words are shown in a separate section, p. 258-278.
Reference should also be made to K. Jacobs' List of serials currently received in the Library of the United States Department of Agriculture (A117), which has a list of abbreviations on p. 331-349. The World list of scientific periodicals (A122) is another authoritative guide for the abbreviation of names of scientific periodicals.

Periodical Lists

A109 BUTTRESS, F. A. Agricultural periodicals of the British Isles, 1681-1900, and their location. Cambridge, Eng., University of

Cambridge, School of Agriculture, 1950. 15 p.
"In general, periodicals relating to botany, economics, forestry, gardening, horticulture, and veterinary science have been excluded."

A110 FRYKHOLM, L. Förteckning över löpande utländska seriepublikationer vid svenska bibliotek inom lantbrukets, veterinärmedicinens, skogshushållningens, jaktens och fiskets ämnesområden. Uppsala, 1950. 69 p.
Mimeographed union list of foreign agricultural journals in Swedish libraries. Compiled by the Librarian of the Royal Swedish Agricultural College at Uppsala.

A111 FRYKHOLM, L. List of specimen copies of foreign agricultural serials, excluding bee and dairy periodicals, in the Library of the Royal Agricultural College of Sweden. Uppsala, 1951. 78 p.

A112 INTERNATIONAL FEDERATION of the AGRICULTURAL PRESS. Repertoire international de la press agricole. Rome, 1939. 419 p.
Directory of 4,091 journals published in 74 countries. Arranged alphabetically by country. Gives place of publication, frequency of issue, and number of copies circulated for each periodical. Has geographic and subject indexes. Preface, foreword, and subject index printed in 5 languages.

A113 INTERNATIONAL INSTITUTE of AGRICULTURE. Catalogue des periodiques possédés par la bibliothèque. Catalogue of periodicals owned by the library. Rome, 1946. 565 p.
French and English. 5,572 entries arranged alphabetically by title with subject index.

A114 PAN AMERICAN UNION. DIVISION of AGRICULTURAL COOPERATION. Tentative directory of agricultural periodicals, societies, experiment stations, and schools in Latin America. Washington, 1945. 90 p.
Arranged by country without descriptive notes or annotations. Periodicals, schools, societies, etc., are given in separate lists under each country.

A115 STUNTZ, S. C., comp. List of agricultural periodicals of the United States and Canada published during the century July, 1810, to July, 1910, ed. by E. B. Hawks. (U.S.D.A. Misc. P. 398) 1941. 190 p.
Alphabetically arranged, entry for each journal being under its latest title with cross reference from other titles. Gives for each periodical: title, frequency of issue, place of publication, dates, and volumes published.

A116 U.S. DEPARTMENT of AGRICULTURE. LIBRARY. List of periodicals currently received in the Library of the United States Department of Agriculture, June 1, 1936, comp. by E. G. Hopper. (U.S.D.A. Misc. P. 245) 1936. 337 p.
----- Supplement, June 1, 1935-December 31, 1940. Washington, 1940. 75 p.
In two parts, the first arranged alphabetically by title and the second by subject. Russian journals are listed in a separate section, but a cross reference to each is included in the alphabetical list. Omits general newspapers and bulletins and circulars of the experiment stations and extension services. Library call numbers are included.

A117 U.S. DEPARTMENT of AGRICULTURE. LIBRARY. List of serials currently received in the Library of the United States Department of Agriculture, November 1, 1949, comp. by Katharine Jacobs. (U.S.D.A. Bibliog. B. 12) 1950. 349 p.
----- Supplement 1, to March 1953, comp. by Ethel L. Coon. 1953. 218 p.
> Complete holdings not listed. Reference to U.S.D.A. Misc. P. 245 (A116) is still required. Has a list of abbreviations on p. 331-349 for words in serial titles in bibliographies prepared in the Library of the U.S. Department of Agriculture.

A118 U.S. DEPARTMENT of AGRICULTURE. LIBRARY. Partial list of United States farm papers received in the U.S. Department of Agriculture Library (State and county farm bureau and state college papers not included). (Libr. List 18, rev.) 1956. 3 p.

A119 U.S. LIBRARY of CONGRESS. REFERENCE DEPARTMENT. SCIENCE DIVISION. Biological sciences serial publications. A world list 1950-1954. Philadelphia, Biological Abstracts, University of Pennsylvania, 1955. 269 p.
> Lists the majority of current biological journals which are of interest to the agricultural scientist.

A120 U.S. LIBRARY of CONGRESS. REFERENCE DEPARTMENT. SCIENCE DIVISION. Scientific and technical serial publications, United States, 1950-1953. Washington, 1954. 238 p.
> A selective list of scientific and technical serial publications issued in the United States between January, 1950, and December, 1953. Pt. I covers serial publications issued by nongovernment sources. Pt. II lists publications issued by federal, state, or local governments, or by organizations and institutions under their sponsorship. Each part has a section listing agricultural serials. This is the second in a series, of which Scientific and technical serial publications, Soviet Union, 1945-1953 was the first.

A121 VARELA, A. and BLANCO, G. Publicaciones periodicas agricolas en America Latina. In Revista Interamericana de Bibliografia, 3(1953):130-137.
> A selected list of Latin American agricultural periodicals arranged by country. Prepared by the Pan American Union, with the help of the U.S. Department of Agriculture Library.

A122 WORLD list of scientific periodicals published in the years 1900-1950. Ed. 3, ed. by William Allan Smith, Francis Lawrence Kent, and George Burder Stratton. New York, Academic Press; London, Butterworths Scientific Publications, 1952. 1058 p.
> Lists approximately 50,000 scientific periodicals and their holdings in British libraries. Important for the identification of periodicals in agriculture and related fields and as a guide to the abbreviation of names of scientific journals.

Societies and Organizations

A123 JUDKINS, J. National associations of the United States. Washington, U.S. Department of Commerce, 1949. 634 p.
> Directory of more than 4,000 associations. Farmers' associations are listed on p. 470-479.

24 A. Agriculture: General

A124 LINDSTROM, D. E. American farmers' and rural organizations. Ed. by H. M. Hamlin. Champaign, Ill., Garrard, 1948. 457 p.
 Textbook for use in colleges. Material is organized into five parts: Pt. 1, The field of farmers' and rural organizations; Pt. 2, Historical backgrounds; Pt. 3, Present-day farmers' organizations; Pt. 4, Comparative structures, functions, and processes; Pt. 5, National policies, rural values, and human welfare. Contains bibliographical footnotes and lists selected readings at the end of each chapter. Indexed.

A125 NATIONAL RESEARCH COUNCIL. Handbook of scientific and technical societies and institutions of the United States and Canada. Ed. 5. (Natl. Res. Council P. 369) 1955. 447 p.
 Includes societies and institutions concerned with agricultural research. A most useful feature is the listing of important serial publications of each group.

A126 UNITED NATIONS EDUCATIONAL, SCIENTIFIC and CULTURAL ORGANIZATION. Directory of international scientific organizations. Paris, 1950. 217 p.
 Includes information concerning agricultural organizations. Gives dates of past meetings and titles of publications.

Tables, Weights and Measures,
Conversion Factors, and Statistical Methods

 General statistical compilations are listed in the agricultural economics section.

A127 FISHER, R. A. and YATES, F. Statistical tables for biological, agricultural, and medical research. Ed. 4, rev. and enl. New York, Hafner, 1953. 126 p.

A128 INTERNATIONAL INSTITUTE of AGRICULTURE. Recueil de coefficients et d'équivalences. Coefficients pour la conversion dans le système métrique décimal des poids, mesures et monnaies; tableaux d'équivalences des unités de mesure du système métrique décimal, en unités de mesures anglo-saxonnes. Ed. 5. Rome, 1937. 293 p.
 Pt. 1, Coefficients for the conversion of weights, measures, and money, is arranged alphabetically by country. Pt. 2 consists of 96 tables showing equivalents of centimeters in inches, feet in meters, bushels per acre in quintaux per hectare, etc. Miscellaneous tables are given in pt. 3 and indexes in pt. 4.

A129 SNEDECOR, G. W. Statistical methods applied to experiments in agriculture and biology. Ed. 5. Ames, Iowa State College Press, 1956. 534 p.

A130 THE STOWAGE red book. A practical hand book for those engaged in production, transportation, stowage, handling, and warehousing of goods. Comp. and ed. by H. H. Hanlin, W. W. Weller, and C. J. Fagg. Ed. 2. New York, Traffic Pub. Co., 1944. 169 p.
 A collection of tables, the principal one arranged alphabetically, and showing various forms of containers used, weight and cubical measurements per unit, and cubic feet to the net ton for approximately 10,500 items of commerce. Other tables give cost per foot

Agriculture; General 25

of space, floor load at various piling heights, floor space, cubic and weight capacity of box cars, etc. Includes information applying to agricultural and food products.

A131 U.S. DEPARTMENT of AGRICULTURE. Weights, measures, and conversion factors. In Agricultural Statistics, 1953, p. vii-x.
Printed in each issue of Agricultural statistics (F19). Contains information from U.S. Department of Agriculture and state schedules concerning the most important agricultural products.

A132 U.S. OFFICE of FOREIGN AGRICULTURAL RELATIONS. Foreign weights and measures with factors for conversion to United States units, by C. G. Gries. Washington, 1944. 23 p.
Includes weights and measures ordinarily used by foreign countries in their agricultural production and trade statistics as well as factors used by the Office of Foreign Agricultural Relations in converting these units to United States equivalents. Tables show common metric units, old Spanish measures of area, etc., with equivalents and units of measure classified by principal countries and territories.

A133 U.S. PRODUCTION and MARKETING ADMINISTRATION. Conversion factors and weights and measures for agricultural commodities and their products. Washington, 1952. 96 p.
"Compiled primarily to provide uniform conversion factors in handbook form for use of Department personnel in converting food and fiber requirements of various segments of the economy into production and allocation units." Arranged by commodities and products. Gives such factors as the relation between live and processed weights of poultry; relation between farm, wholesale, and retail weights of fresh vegetables, etc.; net-content weights of standard containers. Can sizes, domestic and metric measures, and retail weights are given in the appendix.

Classification Systems

A134 INTERNATIONAL INSTITUTE of AGRICULTURE. Systeme de classification des sciences agricoles. Classification scheme of agricultural science. Stoffeinteilung der landwirtschaftswissenschaft, by Sigmund von Frauendorfer. Ed. 2. Rome, 1942. 183 p.
Based on the collection of the Institute's (now F.A.O.) Library in Rome. The decimal classification of the International Institute of Documentation was used as a model. Has a bibliography of general and agricultural classification schemes. The index is in German, French, and English.

A135 U.S. DEPARTMENT of AGRICULTURE. LIBRARY. Classification scheme of the Bibliography of Agriculture, January 1950. (Libr. List 30) 1950. 27 p.
A simple scheme designed especially for the Bibliography of agriculture (A24). No indexes.

A136 U.S. DEPARTMENT of AGRICULTURE. LIBRARY. Scheme of classification for the United States Department of Agriculture Library. Ed. 5. Washington, 1940. 154 p.
Based on a scheme prepared in 1889. Because of inadequacies in

the classification that have developed with the years, the U.S. Department of Agriculture Library does not recommend it for use by any other library. It is still useful, however, for comparative purposes.

SECTION B
PLANT SCIENCES

BOTANY

Botany is one of the most important of the pure sciences as far as agriculture is concerned. Botanical research is a basic part of agronomy, plant breeding, horticulture, forestry, and plant pathology, and its literature must often be used conjointly with theirs. There are, unfortunately, no comprehensive recent guides to botanical reference works. The list here presented includes, in general, only the more important items and those that will be of particular interest to agriculture.

Of the general reference works, one of the most important for a search of botanical literature, particularly for recent articles on applied phases of the science, is Biological abstracts (A11), which has indexed botanical material since the demise of Botanical abstracts (B5). Abstracts are listed in the following units: "Systematic botany," "Morphology and anatomy of vascular plants," "Pharmacognosy and pharmaceutical botany," and "Plant physiology." These are also available in Section D, "Abstracts of Plant sciences," which is published separately. The Bibliography of agriculture (A24) lists articles under systematic and geographic botany, morphology, anatomy and cytology, and physiology. The Agricultural index (A23) lists items under "Botany" and various related subject headings. The Royal Society's Catalogue of scientific papers (C16) is useful for a search of nineteenth-century botanical literature.

Bibliographies of Bibliographies

B1 BAY, J. C. Bibliographies of botany, a contribution toward a bibliotheca bibliographica. In Progressus Rei Botanica 3(1909):331-456.
    An out-of-date but still useful annotated bibliography of bibliographies. Gives detailed information concerning periodicals. Has extensive list of general, local, and subject bibliographies and auction and library catalogs.

B2 BIBLIOGRAPHIES on pure and applied botany and related subjects. (Science Libr. Bibliog. Ser. no. 144) London, 1934. 7 p.

B3 LAWRENCE, G. H. M. Literature of taxonomic botany. In his Taxonomy of vascular plants. New York, Macmillion, 1951. Chap. xiv, p. 284-331.

A succinct and most useful guide to the important literature and reference works used in taxonomic botany.

Abstracting Journals and Reviews

B4 ANNUAL review of plant physiology. v. 1-   1950-   Stanford, Calif., Annual Reviews, 1950-
Surveys current literature in various areas of plant physiology. Contains references.

B5 BOTANICAL abstracts; a monthly serial furnishing abstracts and citations of publications in the international field of botany in its broadest sense. v. 1-15, September 1918-November 1926. Baltimore, Williams & Wilkins, 1918-1926. 15 v.
Merged with Biological abstracts (A11), November 1926. Has classified abstracts with author indexes in v. 1-6, 8-10 and a cumulated author and subject index to the first 10 volumes. V. 11 has a separate author and subject index. V. 12-15 have no indexes. Sections on agronomy, forest botany and forestry, horticulture, and pathology are of particular interest to agriculture.

B6 THE BOTANICAL review . . . v. 1-   January 1935-   Lancaster, Pa., 1935-
Has many review articles with lengthy bibliographies.

B7 BOTANISCHE jahrbücher für systematik pflanzengeschichte und pflanzengeographie. v. 1-   1881-   Leipzig, 1881-1938; Stuttgart, 1939-
Founded by A. Engler. "Literaturbericht" in each issue abstracts current literature. Lack of general index makes it difficult to use as a reference tool.

B8 BOTANISCHES centralblatt; referierendes organ für das gesamtgebiet der botanik. Im auftrage der Deutschen botanischen gesellschaft . . . 1-40 jahrg. (bd. 1-142) 1880-1919; bd. 143-   (neue folge, bd. 1-   ) 1922-   Kassel, Fischer [etc.] 1880-1905; Jena, Fischer, 1906-
Beginning with the new series each volume is in two parts, Referate and Literatur, each part having separate title pages and paging. The Referate section has signed abstracts of important botanical literature published throughout the world. There is a broad subject index and author list for each volume. The Literature section is a list of new literature arranged by broad subjects with an author index.

B9 FORTSCHRITTE der botanik; unter zusammenarbeit mit mehreren fachgenossen. v. 1-   1931-   Berlin, Springer, 1932-
An annual systematic review of international research in the principal branches of botany. Has lengthy bibliographies and annual subject indexes.

Bibliographies and Indexes - Early

B10 JACKSON, B. D. Guide to the literature of botany. Being a classified selection of botanical works, including nearly 6000 titles not given in Pritzel's "Thesaurus." (Index Society Pub. VIII.) London, Pub. for the Index Society by Longmans, 1881. 626 p.

A selective classified subject catalog of short titles listing some 9,000 works. Does not include most elementary works in foreign languages, theses, lectures, inaugural dissertations, works of more medical than botanical interest, and foreign local floras of small areas. The local flora section and the serial publications section, which is arranged geographically, are often useful.

B11 MEISEL, M. A bibliography of American natural history; the pioneer century, 1769-1865. The role played by the scientific societies; scientific journals; natural history museums and botanic gardens; state geological and natural history surveys; federal exploring expeditions in the rise and progress of American botany, geology, mineralogy, paleontology, and zoology. Brooklyn, Premier, 1924-1929. 3 v.

An indispensable aid in the identification of American natural history literature published to 1865. V. 1 is an annotated bibliography with a classified subject and geographic index and a bibliography of biographies. V. 2 and 3 list institutions and their publications. V. 3 also has a chronological list of publications, an index to authors and naturalists, and an index of institutions.

B12 PRITZEL, G. A. Thesaurus literaturae botanicae omnium gentium, inde a rerum botanicarum initiis ad nostra usque tempora, quindecim millia operum recensens. Ed. novam reformatam. Lipsiae, Brockhaus, 1872-[1877] 576 p.

Standard bio-bibliography of early botany and botanists. Has an author list, list of anonymous publications and periodicals, a classified list, and an author index.

Bibliographies and Indexes - Modern

B13 BRITISH MUSEUM (NAT. HIST.) LIBRARY. Catalogue of the books, manuscripts, maps, and drawings in the British Museum (Natural History) . . . London, printed by order of the trustees, 1903-1915. 5 v.
----- Supplement. London, 1922-1940. 3 v.

An author catalog of one of the world's best collections of works on natural history. An indispensable tool for the identification of botanical and other natural history publications.

B14 ENGLER, A. Syllabus der pflanzenfamilien; mit besonderer berücksichtigung der nutzpflanzen nebst einer übersicht über die florengebiete der erde. Ed. 12, completely rev. by Hans Melchior and Erich Werdermann. Berlin-Nikolassee, Borntraeger, 1954-   v. 1-

Bd. 1 - Allgemeiner teil, Bakterien bis gymnospermen. Condenses the Engler system of classification. Has a synoptical key to all families of plants, indicates the various categories to which the families belong, provides synonymies, author citations, and brief family descriptions. Illustrated with diagrams and drawings.

B15 ENGLER, A. and PRANTL, K. Die natürlichen pflanzenfamilien nebst ihren gattungen und wichtigeren arten insbesondere den nutzpflanzen. Ed. 2. Leipzig, Engelmann, 1924-

27 volumes planned; 20 published. Much enlarged over the first edition published in 1887-1909, parts of which are still useful

because second edition is incomplete. Covers the entire plant kingdom with emphasis on useful plants. Arranged taxonomically according to the editors' system. Because of composite authorship there is a lack of uniformity in treatment. Gives concise descriptions with illustrations. Important literature is cited. The standard and indispensable work for systematic botany.

B16   GRAY herbarium card index. Cambridge, Mass.

Issued quarterly to subscribers. Lists on cards all references published since 1873 to new names and new combinations for the flowering plants and pteriodophytes of the Western Hemisphere. About 260,000 cards have been issued to date. The following North American institutions have complete sets of this index: Arnold Arboretum, Carnegie Museum (Pittsburgh), Central Experimental Farm (Ottawa), Chicago Museum of Natural History, Dudley Herbarium (Stanford), Gray Herbarium, Missouri Botanical Garden, Montreal Botanical Garden, New York Botanical Garden, Southern Methodist University, State College of Washington (Pullman), U.S. Department of Agriculture (Beltsville), U.S. National Herbarium (Washington), University of California (Berkeley, Los Angeles), University of Illinois, University of Michigan, University of Minnesota, University of Oklahoma, University of Texas, University of Washington (Seattle), and University of Wisconsin.

B17   HARVARD UNIVERSITY. ARNOLD ARBORETUM. LIBRARY. Catalogue . . . by Ethelyn Maria Tucker . . . Cambridge, Mass., Cosmos Press, 1914-1933. 3 v.

V. 1, Serial publications - Authors and titles; v. 2, Subject catalogue with supplement to v. 1; v. 3, Serial publications - Authors and titles supplement, 1917-1933. Indexes a large collection specializing in dendrology and in general descriptive botany.

B18   INDEX kewensis: an enumeration of the genera and species of flowering plants from the time of Linnaeus to the year 1885 inclusive together with their authors' names, the works in which they were first published, their native countries, and their synonyms. Compiled at the expense of the late Charles Robert Darwin under direction of Joseph D. Hooker by B. Daydon Jackson. Oxford, Clarendon Press, 1895. 2 v.

----- Supplementum . . . no. 1-11 (1886-1950). Oxford, Clarendon press, 1901-1953. 11 parts.

In progress. A monumental and indispensable work used to determine the original publication of a generic name or binomial of a seed plant. Generic names are listed alphabetically and are followed by names of species. Abbreviated references to the author and publication are supplied. Date of publication is indicated occasionally. Country of origin is usually given. B. Daydon Jackson's account of the compilation is in the Journal of the Royal Horticultural Society, 49 (1924):224-229. Supplements 1-3 were also issued as v. 3 of original work.

B19   INDEX londinensis to illustrations of flowering plants, ferns, and fern allies, being an amended and enlarged edition continued up to the end of the year 1920 of Pritzel's Alphabetical register of representations of flowering plants and ferns compiled from botanical and horticultural publications of the XVIIIth and XIXth centuries; prepared under the auspices of the Royal Horticultural Society of London at the Royal Botanic Gardens, Kew, by O. Stapf . . . Oxford, Clarendon Press, 1929-1931. 6 v.

----- Supplement for the years 1921-1935. 1941. 2 v.

"This supplement to the six volumes of the Index londinensis now concludes that work . . . Although no further supplements are to be issued, it has been arranged that references to new illustrations shall be given in the entries of new names included in future supplements of the Index kewensis from 1936 onwards. Such entries will be prefixed by an asterisk."

B20 INTERNATIONAL catalogue of scientific literature: M, Botany. 1st-14th annual issues, 1901-1914. London, Harrison, 1902-1919. 14 v.

The International catalogue of scientific literature is an outgrowth of the Royal Society's Catalogue of scientific papers (C16). Indexes serial material as well as books and monographs. Has an author catalog and a classified subject catalog. Each volume includes list of journals indexed. Indispensable for the period covered.

B21 JUST'S botanischer jahresbericht. Systematisch geordnetes repertorium der botanischen literatur aller länder. Berlin-Zehlendorf, Borntraeger, 1873-1944. 63 v.

A comprehensive classified bibliography of world botanical literature with complete author, subject, and plant name indexes. Useful for the period covered.

B22 KEW. ROYAL BOTANIC GARDENS. LIBRARY. Catalogue. (B. of Misc. Inf. Additional Series III) London, 1899. 790 p.

----- Supplement (1898-1915). London, 1919. 433 p.

In four parts: pt. 1, General; pt. 2, Travels; pt. 3, Periodicals and serials; pt. 4, Manuscripts. Useful for identifying early works.

B23 KROK, T. O. B. N. Bibliotheca botanica suecana ab antiquissimis temporibus ad finem anni mcmxviii. Uppsala and Stockholm, Almquist & Wiksells, 1925. 799 p.

This meticulous bibliography of the works of Swedish botanists is most useful since it includes several very important figures such as Linnaeus. Biographical information is given and books about the botanists are listed. Some information about important non-Swedish botanists is also supplied.

B24 LLOYD LIBRARY, Cincinnati, Ohio. Bibliographical contributions from the Lloyd Library by William Holden and Edith Wycoff. Cincinnati, Lloyd Library, 1911-1919. 3 v.

Principally a bibliography of material in the Lloyd Library relating to botany, exclusive of flora. Alphabetically arranged by author. Also has catalogs of periodical titles and separate lists of material on flora of various parts of the world.

B25 MERRILL, E. D. and WALKER, E. H. A bibliography of Eastern Asiatic botany . . . Sponsored by the Smithsonian Institution, Arnold Arboretum of Harvard University, New York Botanical Garden, Harvard-Yenching Institute. Jamaica Plain, Mass., Arnold Arboretum of Harvard University, 1938. 719 p.

Lists over 21,000 works concerned with plants of Eastern Asia. Alphabetically arranged by author and titles. Has three subject indexes; general, geographic, and systematic. Covers material through 1936. Has excellent list of serial abbreviations which should be useful as a general reference aid since the majority of important botanical journals are noted. The appendix has an extensive list of older oriental works with full bibliography and reference lists of oriental serials.

B26 NEW YORK BOTANICAL GARDEN. Taxonomic index. v. 1-
     1938-   New York, 1938-
     Bimonthly. Issued by the New York Botanical Garden for the American Society of Plant Taxonomists. Aim is to include all current literature in the field of taxonomy and related subjects pertinent to an understanding of the systematics of the plants of the Western Hemisphere. Also available on cards as a part of the card index of the Torrey Botanical Club (B29).

B27 REHDER, A. Bibliography of cultivated trees and shrubs hardy in the cooler temperate regions of the Northern Hemisphere. Jamaica Plain, Mass., Arnold Arboretum of Harvard University, 1949. 825 p.
     Gives references to sources of botanical names, valid names, and synonyms of the woody plants cultivated in the cooler regions of the temperate zone.

B28 REHDER, A. The Bradley bibliography. A guide to the literature of the woody plants of the world published before the beginning of the twentieth century; compiled at the Arnold Arboretum of Harvard University under the direction of Charles Sprague Sargent, by Alfred Rehder. (Pubs. of the Arnold Arboretum no 3) Cambridge, Mass., Riverside Press, 1911-1918. 5 v.
     Titles are arranged systematically according to subjects, and under each separate subject they are arranged chronologically; except periodicals, which are arranged alphabetically. V. 1 includes all botanical publications containing references to woody plants except those restricted to a particular family, genus, or species; these are in v. 2 and are arranged according to the Engler and Prantl system. V. 3 deals with the economic products of woody plants, and with arboriculture, including ornamental uses of trees and shrubs. V. 4 concerns forestry. V. 5 is an author and title index. Serial publications having numerous articles on subjects included in the bibliography are completely analyzed. This carefully done and definitive work is a classic example of what a bibliography should be. During its compilation the author personally examined works in libraries in some 22 countries throughout the world.

B29 TORREY BOTANICAL CLUB. Index to American botanical literature. In Torrey Botanical Club Bulletin, v. 13-   1886-   New York, 1886-
     Since 1894 has also been printed on cards, which are issued monthly. Different subjects as classified may be ordered separately. Purpose is to list "all current botanical literature written by Americans, published in America, or based upon American material; the word America being used to include the entire Western Hemisphere."

B30 U.S. BUREAU of PLANT INDUSTRY. LIBRARY. Botany current literature. Additions to the botanical catalogue, Bureau of Plant Industry, comp. from publications received in the Library of the U.S. Department of Agriculture. v. 1-16, 1919-1934. Washington, 1919-1934.

Continued by Plant science literature (B100), 1935 to 1942 and by Bibliography of agriculture (A24), 1942 to date.

B31 U.S. BUREAU of PLANT INDUSTRY, SOILS, and AGRICULTURAL ENGINEERING. DIVISION of WEED INVESTIGATIONS. Bibliography of weed investigations. 1950-
Issued as an annual in 1950. Quarterly since 1951. The issue for 1950 and the issues for the first two quarters of 1951 were published as monographs in processed form at the Plant Industry Station in Beltsville, Md. Since the July, August, and September issue of 1951 the bibliography has appeared in Weeds, a quarterly journal published by the Weed Society of America. It has also been issued separately as a preprint. Includes references to the economic aspects and general problems of weeds, the botany of weeds, weed control, etc. There are no cumulated author or subject indexes.

B32 VERDOORN, F. Selected references on current research in plant taxonomy, ecology, and geography in Europe, Aftrica, Asia, and Australia. In Chron. Bot. 6(1941):265-287, 298-311.
Arranged alphabetically by author, giving address and nature of research. New World botanists excluded.

Dictionaries

B33 ARTSCHWAGER, E. F. and SMILEY, E. M. Dictionary of botanical equivalents. German-English, Dutch-English, Italian-English by E. Artschwager ... French-English by E. M. Smiley. Baltimore, Williams & Wilkins, 1925. 124 p.

B34 ASHBY, H., and others. German-English botanical terminology; an introduction to German and English terms used in botany, including plant physiology, ecology, genetics, and plant pathology. London, Murby, 1938. 195 p.
H. Ashby, H. Richter, and J. Bärner, joint authors. Concise narrative outline of botanical terminology in German and English on facing pages. Has English and German index; plant names in English, Latin, and German; common names of plant diseases; and list of abbreviations.

B35 BACKER, C. A. Verklarend voordenboek der wetenschappelijke namen van de in Nederland en Nederlandsch-indie in het wild groeiende en in tuinen en parken geweekte varens en hoogere planten. Groningen, Noordhoff, 1946. 664 p.
A dictionary of Latin plant names giving the etymology of generic names and of most binary names, especially those derived from common or personal names.

B36 BAILLON, H. Dictionnaire de botanique. Paris, Hachette, 1876-1892. 4 v.
An extensive and well-illustrated work referring to plant names and botanical terms. Alphabetical arrangement. Has brief bibliographical references. One of the most complete of the botanical dictionaries.

B37 BEDEVIAN, A. K. Illustrated polyglottic dictionary of plant names in Latin, Arabic, Armenian, English, French, German, Italian, and Turkish languages; including economic, medicinal, poisonous, and ornamental plants and common weeds ... Cairo, Argus & Papazian Presses, 1936. 2 pts. in 1 v. 664, 450 p.

B38 BRITTEN, J. and HOLLAND, R. Dictionary of English plant names. London, For the English Dialect Society, Trübner, 1886. 618 p.
    A common-name index with notes about origin and references to the use of the names in literature, often with quotations.

B39 CARPENTER, J. R., comp. An ecological glossary. Norman, University of Oklahoma Press, 1938. 306 p.
    Has definitions of 3,000 terms with references to sources in which terms are used.

B40 DAYTON, W. A. Glossary of botanical terms commonly used in range research. (U.S.D.A. Misc. P. 110, rev.) 1950. 41 p.
    "This glossary has been collated at the request of field officers of the Forest Service primarily for use in connection with the important western floras and other botanical publications which contain no glossaries."

B41 FEATHERLY, H. I. Taxonomic terminology of the higher plants. Ames, Iowa State College Press, 1954. 166 p.
    "About half of the book is given over to a glossary which is ample but intentionally not exhaustive. Obsolete and seldom used words have been omitted purposely." Bibliography, p. 165-166.

B42 GERTH VAN WIJK, H. L. Dictionary of plant names. Published by the Dutch Society of Sciences at Haarlem. The Hague, Nijhoff, 1911-1916. 2 v.
    V. 1 lists Latin names, A-Z. V. 2 is an index of English, French, Dutch, and German names. Gives under Latin name the equivalent popular and literary names in English, Dutch, French, and German. Includes names of wild and cultivated plants, flowers, and fruits; varieties and subvarieties; and parts of plants now or formerly used in medicine or industry.

B43 JACKSON, B. D. Glossary of botanic terms with their derivation and accent. Ed. 4, rev. and enl. Philadelphia, Lippincott, 1928. 481 p.

B44 MARSHALL, W. T. and WOODS, R. S. Glossary of succulent plant terms; a glossary of botanical terms and pronouncing vocabulary of generic and specific names used in connection with zerophytic plants. Pasadena, Calif., Abbey Garden Press, 1945. 112 p.

B45 MARZELL, H. Wörterbuch der deutschen pflanzennamen, mit unterstützung der Preussischen Akademie der Wissenschaften ... Leipzig, Hirzel, 1937-   v. 1-   .
    In progress. Bd. 1-2, Lieferung 5 (A-Helleborus) published by 1956.

B46 STEINMETZ, E. F. Codex vegetabilis. Botanical drugs and spices. Trade-dictionary in 5 languages (Latin, Dutch, German, English, and French) with the botanical origin and with the families. Amsterdam, The Author, 1947. Unpaged.

B47 STEIMNETZ, E. F. Vocabularium botanicum. Ed. 2. Amsterdam, The Author, 1953. 362 p.
    Lists in parallel columns Dutch, German, English, and French equivalents for Latin and Greek botanical terms.

B48 WILLIS, J. C. Dictionary of the flowering plants and ferns. Ed. 6, rev. Cambridge, Eng., Cambridge University Press, 1931. 752 p.
    Includes Latin and popular names and technical terms in one main alphabet.

Directories

B49 CHRONICA botanica. World list, international congresses, societies, etc.; institutions, stations, museums, gardens, societies, and commissions. Special supplement to Chron. Bot. 4(1938):293-383.
 The second part is arranged geographically. A new and more extensive list is in active preparation.

B50 LANJOUW, J. and STAFLEU, F. A. Index herbariorum. Ed. 2. Pt. 1- (Regnum Vegetabile, v. 2.) Utrecht, Netherlands, International Bureau for Plant Taxonomy and Nomenclature of the International Association for Plant Taxonomy, 1954-
 Pt. 1, "The herbaria of the world," lists herbaria alphabetically by location. Contains list of herbarium abbreviations, a list of herbaria by country, and a general index. Pt. 2, "Collectors," lists world collectors and the institutions with which they work. First installment, A-D, issued in 1954. Pt. 3, "Geographical index of collections" and pt. 4, "Authors of types," were projected but not issued by 1956.

B51 WYMAN, D. The arboretums and botanical gardens of North America. In Chron. Bot. 10(1947):395-482.
 Lists arboretums and botanical gardens of the United States, Canada, and Cuba. Has a bibliography. Available also as a reprint.

Flora - Bibliographies

B52 BLAKE, S. F. Guide to popular floras of the United States and Alaska. An annotated, selected list of nontechnical works for the identification of flowers, ferns, and trees. (U.S.D.A. Bibliog. B. 23) 1954. 56 p.
 "In the last half century or so there have appeared hundreds of popular works on plant identification of varying merit and varying degrees of completeness. The present catalog includes the majority of such publications, general, regional, state, and local; the few standard technical manuals for the different regions (but for those for the different states, which will be found listed in S. F. Blake and A. C. Atwood, Geographical guide to floras of the world (B53), pt. 1, 1942); nearly all works that contain illustrations of all or nearly all the species, whether they are technical or popular; and, in addition, a few standard reference books on cultivated plants, weeds, edible plants, useful plants, aquatic plants, plant names, etc."

B53 BLAKE, S. F. and ATWOOD, A. C. Geographical guide to flora of the world, an annotated list with special reference to useful plants and common plant names. Pt. 1- (U.S.D.A. Misc. P. 401) 1942. 336 p.
 Pt. 1: Africa, Australia, North America, South America, and Islands of the Atlantic, Pacific, and Indian Oceans. Pt. 2, to cover other parts of the world, is in preparation. An annotated summary of the available publications on the vascular flora and the economic botany of the world. No list of the scope of this one has ever been published. Includes about 2,597 primary titles and 428 subsidiary titles. Arranged geographically with an author index.

Flora - General

B54 BAILLON, H. Histoire des plantes. Paris, Hachette, 1867-1895. 13 v.
A monumental work treating all families and genera of vascular plants. Many illustrations. Bibliographical footnotes. An English translation of the first 8 volumes, entitled <u>Natural history of plants</u>, was published by Reeve and Co. in London, 1871-1888.

B55 BENTHAM, G. and HOOKER, J. D. Genera plantarum ad exemplaria imprimis in herbariis Kewensibus servata definita. London, Reeve, 1862-1883. 3 v.
Arranged taxonomically with a genera and synonym index for each volume. Descriptions prepared from plants themselves rather than through secondary sources. A monumental and thoroughly documented reference work.

B56 CANDOLLE, A. and C. de, eds. Monographiae phanerogamarum. Paris, Masson, 1879-1896. 9 v.
Is a successor to the <u>Prodromus</u> (B57).

B57 CANDOLLE, A. P., A, and C. de. Prodromus systematis naturalis regni vegetabilis. Paris, Treuttel et Würtz, 1824-1873. 17 v.
Attempts to list all species of seed plants. "While its title of <u>Prodromus</u> (forerunner) designates it merely as a preliminary survey of the vegetable kingdom to be followed by more elaborate monographs, nevertheless it contains the last comprehensive revision of many an important genus." [Stearn, W. T. <u>Candolea</u> 8(1939): 1-4]. General index in last volume.

B58 DALLA TORRE, K. W. von and HARMS, H. Genera siphonogamarum. Leipzig, 1900-1907. 921 p.
Lists spermatophytes with bibliographical references according to the Engler system. Has a complete index. Numbers assigned to genera throughout the work are often used by botanists as indexing guides for the arrangement of material according to the Engler system.

B59 LEMÉE, A. M. V. Dictionnaire descriptif et synonymique des genres de plantes phanérogames. Brest, Imprimerie Commerciale et Administrative, 1929-1951. 8 v.
Designed to include material not listed in older works such as Bentham and Hooker's <u>Genera plantarum</u> (B55) and Engler's <u>Natürlichen pflanzenfamilien</u> (B15). Arrangement in v. 1-7 is alphabetical by Latin name.

B60 NEW YORK BOTANICAL GARDEN. North American flora. New York, 1906-1949. v. 1-34 (incompl.).
In process. Published in parts at irregular intervals. "Designed to present in one work descriptions of all plants growing, independent of cultivation, in North America, here taken to include Greenland, Central America, the Republic of Panama, and the West Indies, except Trinidad, Tobago, and Curaçao and other islands off the north coast of Venezuela, whose flora is essentially South American."

Economic Botany

B61 BOIS, D. Les plantes alimentaires chez tous les peuples et à travers les ages. Histoire, utilisation, culture. (Encyclopédie

Biologique, v. 1, 3, 7, 17) Paris, Lechavalier, 1927-1937. 4 v.
Describes wild and cultivated edible plants. Gives history, uses, vernacular names, and references. V. 1, Vegetables; v. 2, Fruits; v. 3, Spices and condiments; v. 4, Sources of drinks. "May be regarded as the fourth edition of A. Paillieux and D. Bois' Le potager d'un curieux ..., 1885." Detailed index to subjects and illustrations in v. 4.

B62 BURKILL, I. H. A dictionary of the economic products of the Malay Peninsula. London, Crown Agents for the Colonies, 1935. 2 v.
An encyclopedic work of value for information on the economic botany, food, and other products of tropical regions. Arranged alphabetically. Has a native name index.

B63 CLUTE, W. N. The useful plants of the world. Ed. 3. Indianapolis, Clute, 1943. 219 p.
Groups principal species of plants by use and gives their common, technical and family names.

B64 HOLLAND, J. H. Overseas plant products. London, Bale, 1937. 279 p.
List of trade names of "all the natural products of vegetable origin, imported on a commercial scale into the docks under the control of the Port of London Authority and into other ports, for landing and delivery to the consignees in the markets of the United Kingdom. Others of economic value in the countries of production, including many that have been sent to Kew from time to time for identification, are also included." Has bibliography of drugs and medicinal and culinary herbs.

B65 STURTEVANT, E. L. Sturtevant's notes on edible plants. Ed. by U. P. Hedrick. (New York. Department of Agriculture. 27th Annual Report, v. 2, pt. 2; also Report of the New York Agricultural Experiment Station, 1919, II) Albany, 1919. 686 p.
Lists 22,897 plants, giving habitat, uses, principal English names, and references. Edible fungi not included. Has a bibliography. Often useful as a supplement to Index kewensis (B18) in locating information on obscure horticultural plants.

B66 WEHMER, C. Die pflanzenstoffe botanisch-systematisch bearbeitet. Bestandteile und zusammensetzung der einzelnen pflanzenarten und deren produkte. Phanerogamen. Ed. 2. Jena, 1929-1931. 2 v.

----- Ergänzungsband zur zweiten auflage; nachtrage aus den jahren 1930-1934, mitbearb. von Magdalene Hadders. Aus dem nachlass hrsg. von Hans Amelung . . . Jena, Fischer, 1935. 244 p.
Lists systematically 4,493 spermatophytes, with habitat, products, chemical properties, and references.

B67 WIESNER, J. von. Die rohstoffe des pflanzenreichs. Ed. 4. Herausgegeben von Paul Krais und Wilhelm Krais und Wilhelm von Brehmer. Leipzig, Engelmann, 1928-1938. 2 v.
An encyclopedic work emphasizing the chemical features of plants. Has extensive bibliographies and annotated systematic lists of species.

History and Biography

B68 ARBER, Mrs. A. Herbals, their origin and evolution; a chapter in the history of botany, 1470-1670. Ed. 2. Cambridge, Eng.,

Cambridge University Press, 1938. 325 p.
Has a chronological list of the principal herbals and related botanical works published between 1470 and 1670 and an alphabetical list of the historical and critical works consulted during the preparation of the work.

B69   BRITTEN, J. and BOULGER, G. S. A biographical index of deceased British and Irish botanists. Ed. 2. London, Taylor & Francis, 1931. 342 p.
Useful for bibliographical searching since reference is made to botanical works of the botanists listed. References are also given to the chief sources of further information.

B70   GREEN, J. R. A history of botany, 1860-1900. Oxford, Clarendon Press, 1909. 543 p.
A continuation of Sachs' History of botany, 1530-1860 (B72).

B71   REED, H. S. A short history of the plant sciences. Waltham, Mass., Chronica Botanica, 1942. 320 p.
"References" at end of each chapter except first.

B72   SACHS, J. von. History of botany, 1530-1860. Tr. by Henry E. F. Garnsey, rev. by Isaac Payley Balfour. Oxford, Clarendon Press, 1906. 568 p.

B73   VERDOORN, F. and J. G. The index botanicorum. Waltham, Mass., Chronica Botanica. [In preparation, 1957.]
This is designed to be a critical biographical dictionary of plant scientists of all times. It is being prepared under the editorship of F. and J. G. Verdoorn under the auspices of the International Biohistorical Commission, with the assistance of the Arnold Arboretum of Harvard University [See: Chron. Bot. 8(1944):425-448]. Before publishing any installments of the Index botanicorum, the Commission proposed to issue a three volume concise Biographical dictionary of plant scientists, which is now in active preparation.

Miscellaneous

B74   HITCHCOCK, A. S. Manual of the grasses of the United States. Ed. 2, rev. by Agnes Chase. (U.S.D.A. Misc. P. 200) 1950. 1051 p.
Standard reference work on use, distribution, and botany of grasses in the continental United States, excluding Alaska. Describes 169 numbered genera and 1,398 numbered species, nearly all of which are illustrated. Based mainly on material in the U.S. National Herbarium. Includes distribution maps, a synonymy, a glossary, and an index. Bibliography is based on the catalog of grasses maintained in the Division of Grasses.

B75   INTERNATIONAL CONGRESS of BOTANY. 8th, Paris, 1954. International code of botanical nomenclature, prep. by J. Lanjouw and others. Utrecht, International Bureau of Plant Taxonomy and Nomenclature, 1956. 338 p.

B76   MUENSCHER, W. C. L. Poisonous plants of the United States. New York, Macmillan, 1951. 277 p.
Literature references: p. 246-260.

B77   MUENSCHER, W. C. L. Weeds. New York, Macmillan, 1955. 560 p.
A treatise on dissemination and control of weeds followed by a

systematic arrangement of families (Engler classification), a dichotomous key to the species, and descriptions and illustrations of the species of weeds of Northern United States and Canada.

B78   U.S. FOREST SERVICE. Range plant handbook. Washington, 1937. 512 p.
Evaluates 300 or more of the outstanding "key" plants of western ranges as regards grazing, watershed protective cover, recreational, and other uses.

B79   VAN DERSAL, W. R. Native woody plants of the United States, their erosion control and wild life values. (U.S.D.A. Misc. P. 303, rev.) 1939. 362 p.
Indexed list, arranged alphabetically by Latin name. Maps showing soil and plant growth regions and climatic provinces are included. Bibliography of 649 references and list of common names of woody plants are appended.

B80   WEINTRAUB, F. C. Grasses introduced into the United States. (U.S.D.A. Agr. Handb. 58) 1953. 79 p.
An annotated list with bibliography. Has a common name index.

B81   WODEHOUSE, R. P. Hay fever plants. Waltham, Mass., Chronica Botanica, 1935. 245 p.
Arranged taxonomically. Has regional surveys, glossary, and bibliography.

HORTICULTURE AND AGRONOMY

Reference material on horticultural and field crops is included in this section. It is difficult to separate horticultural works from materials concerning botany and agronomy since they are so closely interrelated. There is also much confusion as to whether certain vegetable crops should be considered as field or horticultural crops. This confusion is reflected in the literature. Horticultural abstracts (B83), for instance, recently announced that "in the current year, 1952, potatoes and sweet potatoes will be dealt with both in Horticultural and in Field crop abstracts [B84], but as from 1953 they will pass from our care to that of Field crop abstracts as being nowadays almost always grown as field crops." "Agronomy" is a somewhat ambiguous term which has been used in various ways throughout its career. In its broadest sense it covers the whole field of agriculture, but it is used here as applying principally to the cultivation of field and herbage crops (cereals, legumes, root crops, fiber crops, oil crops, grasses and herbage crops, seeds and seed production, and related subjects).

Works listed under Botany and Plant Pathology will, of course, be of interest to the horticulturists and agronomists. General agricultural reference publications will also be useful. The Agricultural index (A23), for instance, indexes several general horticultural and agronomic journals in the English language. The Experiment station record (A22) had a section entitled "Horticulture" which listed items about vegetable and truck crops as well as fruits, although it did not treat ornamental horticulture. It also had a section on field crops in each issue. The Bibliography of agriculture (A24) under a main heading "Plant Science" lists material on field and horticultural crops. Biological abstracts (A11) usually has monthly sections on horticulture and agronomy: the separately published Section D, "Abstracts of plant sciences," includes abstracts on horticulture and agronomy. <u>Agricultural and horticultural engineering abstracts</u> (D55), as its title suggests, has a section of interest to horticulturists.

Abstracting Journals and Annual Reviews

B82   ADVANCES in agronomy, ed. by A. G. Norman. v. 1-   1949-
   New York, Academic Press, 1949-
   Reviews progress in basic research in soil and crop science and developments in agronomic practice. Contains references. Issued under the auspices of the American Society of Agronomy.

B83   COMMONWEALTH BUREAU of HORTICULTURE and PLANTATION CROPS. Horticultural abstracts. v. 1-   March 1931-
   East Malling, Eng., 1931-
   Gives world coverage. "For the purposes of this journal horticultural crops include all those whose products normally appear as fruits, vegetables, nuts, and flowers, while all tropical and subtropical perennials, such as tea, rubber, oil palm, etc., are considered to be plantation crops." Each volume has separate subject and name indexes. Issued by the Imperial Bureau of Fruit Production, 1931-1937; Imperial Bureau of Horticulture and Plantation Crops, 1938-1948.

B84   COMMONWEALTH BUREAU of PASTURES and FIELD CROPS. Field crop abstracts. v. 1-   January 1948-   Aberystwyth, Wales, 1948-
   Issued quarterly. Covers world literature on the cultivation and agricultural botany of crops grown in rotation and includes sections on land utilization and management, machinery and equipment, conservation, and the economics of crop production. Each issue has an author index. Annual volumes are indexed by author, subject, genera, crop variety, diseases, pests and parasites, and geographical names.

B85   COMMONWEALTH BUREAU of PASTURES and FIELD CROPS. Herbage abstracts. v. 1-   June 1931-   Aberystwyth, Wales, 1931-
   Issued quarterly. Supersedes <u>Current literature lists</u>, <u>Herbage research circulars</u>, and <u>Monthly abstract service</u>. Abstracts material on herbage and fodder plants published in more than 60 countries. Has author and genera indexes in each recent issue. The 1951 volume has an author index and a subject index which includes generic names. Earlier volumes usually have separate author, genera, disease and pest, geographical names, and subject indexes. Formerly issued by Imperial Bureau of Pastures and Forage Crops and by the Imperial Bureau of Plant Genetics.

B86   DIE GARTENBAUWISSENSCHAFT. v. 1-   1928-   Berlin, 1929-
   Issued in two sections: "Originalien" and "Referate." The "Referate" section, which is paged separately, has lengthy abstracts of important horticultural articles in world literature.

Bibliographies and Indexes

Since the bibliographies of specific crops are so numerous, they are not listed here. However, the annual indexes of the <u>Bibliography of</u>

agriculture (A24) and the Experiment station record (A22) list crop bibliographies under the general heading "Bibliographies" and under the name of the plant. (See Section G, Addendum, for additional material.)

B87 CATON, D. Selected bibliography - range resources and management. Moscow, Idaho, University of Idaho, 1954. Unpaged.
A mimeographed list, arranged alphabetically by author, of references in range management published in American periodicals, bulletins, and other sources.

B88 FRANCK, W. J. and BRUIJNING, W. H. General seed bibliography. Wageningen, 1931. 632 p.
A comprehensive list first published in 1928 in mimeographed form as Bibliography of "Germination of seed."

B89 HARVEY, R. B. An annotated bibliography of the low temperature relations of plants. Minneapolis, Burgess, 1935. 223 p.
A comprehensive list of references from world literature.

B90 INTERNATIONAL SEED TESTING ASSOCIATION. Proceedings. Copenhagen, 1925-
Issued irregularly. Each issue has a title bibliography of world publications on seed testing and seed control.

B91 MASSACHUSETTS HORTICULTURAL SOCIETY. LIBRARY. Catalogue. Cambridge, Mass., 1918. 587 p.
A well-made author and subject catalog of approximately 22,000 volumes. Useful for identification of book and periodical references to early horticultural literature.

B92 MOLON, G. Bibliografia orticola. (Stazione Sperimentale di Orti-Frutticoltura. B. 3) Milano, Terragni & Calegari, 1927. 428 p.
Covers horticulture in general with emphasis on references to the literature of viticulture. Useful occasionally as a guide to European material.

B93 PENNSYLVANIA HORTICULTURAL SOCIETY. LIBRARY. Catalog. Philadelphia, 1941. 161 p.

B94 PIETERS, A. J. A digest of pasture research literature in the continental United States and Canada, 1885 to 1935. Washington, Bureau of Plant Industry, Division of Forage Crops and Diseases, 1936. 130 p.
----- Supplement. Washington, 1937. 12 p.
Arranged by state with subject index. Selective. No range literature is included. Supplement covers literature up to February 1937.

B95 PIETERS, A. J. A digest of some world pasture research literature (exclusive of the continental United States and Canada). Washington, Bureau of Plant Industry, Division of Forage Crops and Diseases, 1937. Unpaged.
Supplements B94. Arranged by country with subject index.

B96 ROCKWELL, J. E. Contents of and index to bulletins of the Bureau of Plant Industry, nos. 1 to 100 inclusive. (U.S. Bureau of Plant Industry B. 101) 1907. 102 p.

B97 ROYAL HORTICULTURAL SOCIETY, London. LIBRARY. The Lindley Library; catalogue of books, pamphlets, manuscripts, and drawings. London, 1927. 487 p.
A catalog of one of the world's largest horticultural collections. Majority of entries arranged alphabetically under author. Articles in periodicals, government publications, and publications of

42   B. Plant Sciences

societies are listed. Pamphlets, dissertations, and anonymous works are included.

B98   U.S. BUREAU of PLANT INDUSTRY. LIBRARY. Agronomy current literature. v. 1-v. 9, no. 26; January 1926-December 1934. Washington, 1926-1934.

Absorbed into Plant science literature (B100). Essentially a checklist; useful only occasionally in a literature search. References published during the period can usually be found in the Experiment station record (A22) or other sources.

B99   U.S. BUREAU of PLANT INDUSTRY. LIBRARY. Check list of publications issued by the Bureau ... 1901-1920, and by the divisions and offices which combined to form this Bureau, 1862-1901. (U.S.D.A. Libr. Bibliog. Contrib. 3) 1921. 127 p.

B100 U.S. BUREAU of PLANT INDUSTRY. LIBRARY. Plant science literature; selected references comp. by the Library staff ... (mainly for the Botanical catalog) from publications received in the U.S. Department of Agriculture Library ... Washington, 1935-1942. 15 v.

Formed by the union of Agronomy current literature (B98) and Botany current literature (B30). Superseded by the Bibliography of agriculture (A24).

B101 U.S. DEPARTMENT of AGRICULTURE. LIBRARY. Cotton literature; selected references. Washington, 1931-1942. 12 v.

Continued by the Bibliography of agriculture (A24). A comprehensive list useful for years covered.

Dictionaries, Encyclopedias, and Manuals

See Section G, Addendum, for additional material.

B102 AMERICAN JOINT COMMITTEE on HORTICULTURAL NOMENCLATURE. Standardized plant names. Ed. 2, a rev. and enl. listing of approved scientific and common names of plants and plant products in American commerce or use. Harrisburg, Pa., McFarland, 1942. 675 p.

The purpose of this valuable reference tool is to standardize names in use in the horticultural trade. Particularly useful for its list of horticultural varietal names and its special lists such as Drug plant names, Fruit and edible nut names, Economic plants, Plant patents, Lumber trade names, and Nut group common names. Several of the lists of interest to agriculture, such as those concerning the cereals and grasses, were prepared by the Bureau of Plant Industry.

B103 BAILEY, L. H. Manual of cultivated plants most commonly grown in the continental United States and Canada. Rev. ed., completely restudied. New York, Macmillan, 1949. 1116 p.

Arranged systematically with indexes to scientific and common names. Has a glossary and key to families.

B104 BAILEY, L. H. Standard cyclopedia of horticulture. New York, Macmillan, 1914-1917. 6 v. 1947 reissue in 3 v.

Purpose is to describe "the plants horticulturally grown within its territory" (North America, Porto Rico, Hawaii, and the other islands) to the close of the year 1912, discuss growing practices, and to "depict the horticultural capabilities of the states and provinces." Based on his Cyclopedia of American horticulture but is largely a new work. In spite of "editions" dated as late as 1949 there have been no significant revisions since the original edition of 1917. A "second edition" issued in 1922 included only the correction of typographical errors. All later issues or "editions" are reprints of the 1922 version. Articles are by specialists or the editor. "It is intended to account for all the species in the trade," but names of garden varieties are given only incidentally. Useful for botanic as well as cultural information. Material on American native plants, shrubs, and trees is particularly extensive. References to illustrations, both colored and black and white, in botanical literature are given. "A synopsis of the plant kingdom," by Karl M. Wiegand; "Key to the families and genera," "English equivalents of Latin names of species," and "Glossary of botanical terms" are useful addenda in v. 1. The supplement in v. 6 has a list of collaborators, a "Cultivator's guide to the articles," "Additional species," a "Finding list of binomials," and an "Index to synonyms, vernacular names, and miscellaneous references, not in alphabetic order in the Cyclopedia."

A comprehensive reference manual still useful in spite of its age.

B105 BAILEY, L. H. and E. Z. Hortus second; a concise dictionary of gardening, general horticulture, and cultivated plants in North America. New York, Macmillan, 1941. 778 p.

Designed to include plants in cultivation in North America to mid-year 1940. In effect a supplement to and condensed version of the Standard cyclopedia of horticulture (B104). The inventory is based on catalogs of plant and seed dealers, current literature, herbaria, correspondence, etc. Common names, descriptive terms and definitions, inventories of families and plants, and abundant cross references are included. "The first Hortus was published in December 1930 . . . for the period ending 1929. The old book now becomes an historic record, and should be kept in the library as a reference."

B106 CHITTENDEN, F. J., ed. Dictionary of gardening. Oxford, Oxford University Press, 1951. 4 v.
----- Supplement. 1956. 334 p.

A definitive work prepared under the auspices of the Royal Horticultural Society. Partly based on Nicholson's Dictionary of gardening. The first four volumes "deal with species and with hybrids and forms that are long-established and widely grown." The supplement refers to new plants introduced during the preparation of the Dictionary. It also contains lists of recommended varieties improved by hybridization and selection.

B107 HEDRICK, U. P. Cyclopedia of hardy fruits. Ed. 2, enl. New York, Macmillan, 1938. 402 p.

Describes varieties of hardy fruits grown in North America. Has glossary and indexes to species and synonyms. The following works by Hedrick are also useful: Cherries of New York (Albany, 1915), Grapes of New York (Albany, 1908), Peaches of New York (Albany, 1916), Pears of New York (Albany, 1921), Plums of New York (Albany, 1911), Small fruits of New York (Albany, 1925).

B108 HEDRICK, U. P. The vegetables of New York. (Report of the New York State Agricultural Experiment Station for the year[s]

ending June 30, 1928, June 30, 1931, June 30, 1934, June 30, 1935.) Albany, Lyon Printers, 1928-1937. V. 1, pts. 1-4.
Contents: v. 1, pt. 1, Peas of New York; v. 1, pt. 2, Beans of New York; v. 1, pt. 3, Sweet corn; v. 1, pt. 4, Cucurbits. Each part has extensive lists of references and an index.

B109 TAYLOR'S encyclopedia of gardening, horticulture, and landscape design. Ed. by Norman Taylor. Ed. 2. Boston, American Garden Guild and Houghton Mifflin, 1948. 1225 p.
First issued in 1936 as The garden dictionary. An excellent popular compilation. Articles are by specialists but written for the layman. Plants are listed under scientific name with cross references from the popular name. Includes information on gardening conditions in various geographical divisions, types and methods of gardening, methods of propagation, etc.

B110 ZANDER, R. and HECKEL, M. Wörterbuch der gärtnerischen fachausdrücke in vier sprachen. Dictionary of horticultural terms. Berlin, Lang, 1938. 419 p.
Compiled in connection with the preparatory work for the 12th International Horticultural Congress. Restricted to words frequently appearing in the Congress reports. In 4 parts: German, English, French, and Italian. Terms are given in each language with equivalents in the other three.

Directories and Lists

See note on directories, p. 15.

B111 ANNUAIRE fructidor . . . annuaire international des fruits, legumes, primeurs, dérivés et industries annexes, 1952. Ed. 17, Paris, 1952. 1736 p.
Fruit and vegetable directory showing centers of production in France and French colonies and containing lists of railroad and shipping companies, shippers, exporters and importers in France, French colonies, and foreign countries.

B112 BROOKS, R. M. and OLMO, H. P. Register of new fruit and nut varieties, 1920-1950. Berkeley, University of California Press, 1952. 206 p.
Lists 1,106 varieties of fruit and nuts. Has a complete list of patented varieties. Based on material issued in the Proceedings of the American Society for Horticultural Science from 1944 through 1950.

B113 FRUIT annual, 1952-1953: year-book and directory of the world's fruit trade. London, British-Continental Trade Press, 1953. 423 p.
Contains directories of fruit producers and traders throughout the world.

B114 GRAIN trade buyers' guide, 1954. no. 10. Chicago, Grain and Feed Journals Consolidated, 1954. 55 p.
Issued annually. Contains classified directory of equipment, supplies, services, and wholesale field-seed processors.

B115 THE HORTICULTURAL guide, 1954. New York, 1954. 184 p.
A directory of nurseries and florists in New York, Connecticut, and New Jersey.

Horticulture and Agronomy 45

B116 HORTICULTURAL trade directory. Ed. 8, 1940-1942. New York, De La Mare, 1940. 552 p.
Contains a classified list of firms and individuals engaged in the florist, nursery, and seed industries, arranged by states and cities in the United States. Lists of horticultural trade organizations, firms issuing catalogs, cemetery superintendents, park executives, and landscape architects are also included. Federal plant quarantines, and lists of plant patents and state flowers are also given.

B117 IRVING, A. J., ed. The directory of American horticulture for 1954. Ithaca, N. Y., American Horticultural Council, 1953. 96 p.
Lists organizations in the United States and Canada, universities and colleges teaching horticulture, trial grounds and test gardens, all-America rose and seed selections, botanic gardens and arboretums, public and other gardens, garden centers, state flowers and trees, and horticultural awards.

B118 NATIONAL fruit and vegetable directory, 1951-1952: comp. and produced for the fruit, vegetable, and allied industries. Chicago, Aberdeen Press, 1951. 208 p.
Issued annually. A geographic listing of dealers in the United States and Canada.

B119 NATIONAL HAY ASSOCIATION. Year book, 1952-1953. Indianapolis, 1952. 117 p.
Contains a membership directory and other information concerning the hay trade.

B120 RICE annual, 1954. New Orleans, Rice Journal, 1954. 92 p.
Contains statistics and directories of rice mills and rice driers.

B121 SEED trade buyers' guide, 1954. V. 37. Chicago, Seed World, 1954. 284 p.
Published annually. Lists federal and state seed laws, foreign seed firms, seed statistics, organizations and societies, and suppliers of equipment and supplies. A valuable reference tool.

B122 SMALLWOOD, N. W. Commercial horticultural organizations of the United States and Canada. (U.S.D.A. Libr. Libr. List 17, rev.) 1952. 8 p.
Arranged geographically.

B123 SMALLWOOD, N. W. Horticultural organizations of the United States and Canada. (U.S.D.A. Libr. Libr. List 16, rev.) 1952. 13 p.
Arranged geographically.

B124 STEFFEK, E. F., ed. Plant buyers' guide of seed and plant materials in the trade. Ed. 5, rev. Boston, Horticultural Society, 1949. 259 p.
List of plants arranged by scientific name with key to nurserymen and seedmen offering them.

Handbooks and Texts

B125 MARTIN, J. H. and LEONARD, W. H. Principles of field crop production. New York, Macmillan, 1949. 1,176 p.
Designed as a text but useful as a reference source because of selected bibliographies concerning the various crops. Has a glossary.

B126 WHEELER, W. A. Forage and pasture crops, a handbook of information about the grasses and legumes grown for forage in the

United States. New York, Van Nostrand, 1950. 752 p.
Has selected references on forage crops arranged by states.

History

B127 HEDRICK, U. P. A history of horticulture in America to 1860. New York, Oxford University Press, 1950. 551 p.
The chapter on "Horticultural literature, 1700-1860" is particularly useful for descriptive material on early American horticultural works and periodicals. The chapter on horticultural societies and the bibliography also have much reference value.

Periodical Lists

B128 SHEPARD, S. Bibliography of current horticultural periodicals, 1954. Ithaca, N. Y., Bailey Hortorium of the New York State College of Agriculture, 1955. 54 p.
A list, arranged by country, of currently published horticultural journals. Also serves in part as a general list of agricultural periodicals, the term horticulture being liberally interpreted to include some periodicals devoted to agriculture, forestry, and park management. The so-called household magazines with gardening sections are not included. Has a subject index.

PLANT BREEDING

Plant breeding, the science of genetics applied to plants, is one of the more important fields of agricultural research. Although there are several genetics journals, such as the Journal of genetics, Heredity, and Genetics, much of the literature is widely dispersed throughout agricultural and general scientific publications. In making a literature search for material on plant breeding the geneticist must examine the general indexes as well as those listed in this section. The Bibliography of agriculture (A24), for instance, lists publications on plant breeding in the "Plant science" section under the heading "Genetics." The genetics section of Biological abstracts (A11) includes material on both animal and plant breeding. The Agricultural index (A23) lists items under "Plant breeding" and other subject headings. The Experiment station record's (A22) section on genetics also includes material on plant and animal breeding.

The plant geneticist will also find useful several of the referecne works listed in the Horticulture and Agronomy, Botany, and Animal sciences sections.

Abstracting Journals

B129 COMMONWEALTH BUREAU of PLANT BREEDING and GENETICS. Plant breeding abstracts. v. 1-   January/June 1930-   Cambridge, Eng., 1930-
Published quarterly. V. 1-8 issued by Imperial Bureau of Plant Genetics. Covers world literature on plant breeding. Has lengthy

abstracts of the more important items. Author index for each issue. Annual indexes cover subjects and authors. Before 1950 items were classified by the International decimal system.

B130 RESUMPTIO genetica. v. 1-19. 1925-1953. The Hague, Nijhoff, 1926-1953. 19 v.

Each volume lists new literature on genetics and abstracts the most important items. Has author, subject, and Latin name indexes.

B131 ZEITSCHRIFT für pflanzenzüchtung. v. 1-   1912-   Berlin, Parey, 1913-

Issued quarterly. Has original articles plus lengthy abstracts of important items. Annual author and subject indexes.

B132 DER ZÜCHTER; zeitschrift fur theoretische und angewandte genetik ... v. 1-   1929-   Berlin, 1929-

Issued 8 times a year. Majority of articles are on plant genetics. Has abstracts and book reviews of recent literature in each issue. Annual author and subject indexes.

Bibliographies

B133 BIBLIOGRAPHIA genetica. v. 1-   1925-   's-Gravenhage, Nijhoff, 1925-

Issued irregularly. Each volume contains lengthy monographs on separate subjects. Of reference value because of the excellent bibliographies.

B134 IMPERIAL BUREAU of PLANT BREEDING and GENETICS. Bibliography on cold resistance in plants. Cambridge, Eng., 1939. 22 p.

B135 IMPERIAL BUREAU of PLANT GENETICS. Barley breeding bibliography. Cambridge, Eng., 1931. 24 p.

B136 IMPERIAL BUREAU of PLANT GENETICS. Bibliography on the breeding and genetics of the millets and sorghums. Cambridge, Eng., 1932. 21 p.

B137 IMPERIAL BUREAU of PLANT GENETICS. Breeding resistant varieties, 1930-1933. Cambridge, Eng., 1935. 32 p.

A supplement to Breeding varieties resistant to disease (B138). Comprises papers summarized in Plant breeding abstracts (B120).

B138 IMPERIAL BUREAU of PLANT GENETICS. Breeding varieties resistant to disease. Cambridge, Eng., 1930. 43 p.

Papers on this subject after 1930 are listed in Plant breeding abstracts (B129) and in Breeding resistant varieties, 1930-1933 (B137).

B139 IMPERIAL BUREAU of PLANT GENETICS. The experimental production of haploids and polyploids. Cambridge, Eng., 1936. 28 p.

B140 IMPERIAL BUREAU of PLANT GENETICS. Interspecific and intergeneric hybridization in relation to plant-breeding. Cambridge, Eng., 1932. 30 p.

B141 IMPERIAL BUREAU of PLANT GENETICS. Oat breeding bibliography. Cambridge, Eng., 1931. 21 p.

B142 IMPERIAL BUREAU of PLANT GENETICS. Rice breeding bibliography. Cambridge, Eng., 1932. 26 p.

B143 IMPERIAL BUREAU of PLANT GENETICS. Rye breeding bibliography. Cambridge, Eng., 1932. 26 p.
B144 IMPERIAL BUREAU of PLANT GENETICS. Wheat breeding bibliography. Cambridge, Eng., 1931. 3 v.
B145 IMPERIAL BUREAU of PLANT GENETICS. SCHOOL of AGRICULTURE. Tobacco breeding bibliography. Cambridge, Eng., 1937.
B146 MATSUURA, H. A bibliographic monograph on plant genetics (genic analysis) 1900-1929. Ed. 2, rev. Sapporo, Hokkaido Imperial University, 1933. 787 p.

Pt. 1, Genic analysis of plants. (arranged alphabetically by plant name with abstract of work done); pt. 2, Bibliography; pt. 3, Author and subject indexes.

B147 STUBBE, H. Genetisch-pflanzenzüchterische bibliographie, 1939-1946 (1947). (Der Züchter. Sonderheft [1] 1948.) Berlin, Springer, 1949. 287 p.

A survey of references to world research in plant breeding during the war years. Arranged by subject. No author index.

B148 WARNER, M. F., and others. A bibliography of plant genetics. (U.S. D.A. Misc. P. 164) Washington, 1934. 552 p.

Dictionaries

B149 KNIGHT, R. L. Dictionary of genetics, including terms used in cytology, animal breeding, and evolution. (Lotsya, a Biological Miscellany, v. 2.) Waltham, Mass., Chronica Botanica, 1948. 183 p.
B150 RIEGER, R. and MICHAELIS, A. Genetisches and cytogenetisches wörterbuch. (Der Zuchter. Sonderheft 2, 1954.) Berlin, Springer, 1954. 140 p.

A comprehensive dictionary of genetical terms which should be useful as a supplement to Knight (B149).

Directories

B151 FOOD and AGRICULTURE ORGANIZATION of the UNITED NATIONS. List of plant breeders in Canada and the United States of America; a reference list of classified information giving the addresses and activities of plant breeders in these countries. Washington, 1949. 64 p.

Lists breeders in official institutions only.

B152 FOOD and AGRICULTURE ORGANIZATION of the UNITED NATIONS. World list of plant breeders. Rome, 1952. 105 p.
Supplements B151.
B153 INTERNATIONAL UNION of BIOLOGICAL SCIENCES. Index des généticiens. Paris, 1953. 113 p.

An index of world geneticists arranged alphabetically by name and geographically by institution.

Handbooks and Texts

B154 BABCOCK, E. B. and CLAUSEN, R. E. Genetics in relation to agriculture. New York, McGraw-Hill, 1927. 673 p.
List of literature, p. 613-647.
B155 DARLINGTON, C. D. and AMMAL, E. K. J. Chromosome atlas of cultivated plants. London, Allen & Unwin, 1945. 397 p.
Bibliography, p. 355-382. Has index to families and genera.
B156 HAYES, H. K., and others. Methods of plant breeding. Ed. 2. New York, McGraw-Hill, 1955. 551 p.
Literature citations, p. 497-518. Glossary, p. 519-526.

PLANT PATHOLOGY

"Plant pathology is that phase of botanical science which deals with the diseases or troubles of plants." (Heald, F. D., Manual of plant diseases, 1933, p. 1). Another term used for the science of plant pathology is phytopathology. The principal types of diseases with which plant pathologists are concerned are: nonparasitic diseases related to the physiological processes of the plant, bacterial diseases, virus diseases, and diseases caused by fungi. Troubles caused by animal parasites such as the nematodes, protozoa, and insects are also of interest to the plant pathologist although the economic entomologist is the specialist in the fight against insects.

The literature of plant pathology is large and much of it appears in botanical, general science, and medical publications. As G. W. Fischer states, "Our accumulated knowledge in the field of mycology and plant pathology is scattered through literally hundreds of series of journals, periodicals, proceedings, archives, reports, bulletins, books, etc., representing many thousands of volumes and in some cases dating back more than a hundred years. The problem is how to get in touch with the literature without spending a disproportionate amount of time at the task even with the aid of such abstracting journals as the Review of applied mycology [B160]." (The smut fungi ... (B193), 1950, p. v.).

The general indexes and abstracting journals which are described in the "Agriculture:General" section must be used by the plant pathologist. The Agricultural index (A23), for instance, lists references under such headings as "Plants - Diseases and pests," "Fungi in agriculture," "Deficiency diseases in plants," etc., and is useful as a finding guide before the item wanted is abstracted in Biological abstracts (A11) or the Review of applied mycology (B160). In the Bibliography of agriculture (A24), references are listed under "Plant science - Pathology" and are separated as follows: "General," "Diseases caused by fungi," "Viruses and virus diseases," "Diseases caused by bacteria," and "Diseases caused by nematodes." Although no longer published, the Experiment station record's (A22) section on "Diseases of plants" should not be forgotten. Biological abstracts separates its references on phytopathology as follows: "Diseases caused by fungi," "Diseases caused by bacteria," "Diseases caused by animal parasites," "Virus diseases," "Non-parasitic diseases," "Parasitism and resistance," "Disease control," and "Miscellaneous." Plant pathology abstracts, beginning in 1939, have been included in Section D, "Abstracts of plant sciences," which is available separately as a reprint. Other reprint sections of Biological abstracts—such as Section E, which includes abstracts on parasitology, protozoology, and economic entomology—would be of interest to the plant pathologist.

There are several texts such as F. D. Heald's <u>Manual of plant diseases</u> (New York, McGraw-Hill, 1933), J. G. Dickson's <u>Diseases of field crops</u> (New York, McGraw-Hill, 1947), and Sir Edwin J. Butler's <u>Plant pathology</u> (London, Macmillan, 1949), which have well-selected lists of references and other basic information.

General Works

B157 U.S. AGRICULTURAL RESEARCH SERVICE. HORTICULTURAL CROPS RESEARCH BRANCH. Plant disease reporter. v. 1- August 15, 1917-   Washington, 1917-
----- Supplement. no. 1-   May 15, 1919-   Washington, 1919-
Monthly. Title was <u>Plant disease bulletin</u> from August 1917 to December 1923. Contains reports, summaries, observations, and comments submitted voluntarily by qualified observers. Reference value enhanced by indexes for each volume. The <u>Supplement</u> usually publishes the more lengthy papers.

Abstracting Journals

B158 ABSTRACTS of bacteriology. v. 1-9, February 1917-December 1925. Baltimore, Williams & Wilkins, 1917-1926. 9 v. illus.
Bimonthly, 1917-1920; monthly, 1921-1925. United with <u>Botanical abstracts</u> (B5) to form <u>Biological abstracts</u> (A11). Abstracted world literature in bacteriology, mycology, protozoology, including such specific subjects as industrial bacteriology, mycology, soil bacteriology, plant bacteriology, dairy bacteriology, parasitology, medical entomology, etc. Annual author and subject indexes.

B159 ANNALES mycologici editi in notitiam scientiae mycologicae universalis. Ed. by H. Sydow. v. 1-41, 1903-1943. Berlin, Friedlaender and Sohn, 1903-1943.
Has original papers relating to mycology, abstracts, book reviews, and a new-literature section. Name indexes were issued for most volumes. Continued by <u>Sydowia annales mycologici editi in notitiam scientiae mycologicae universalis</u>, ed. by F. Petrak. (v. 1-   1947- Horn, N.Ö., Austria, 1947-   ).

B160 COMMONWEALTH MYCOLOGICAL INSTITUTE. The review of applied mycology. v. 1-   January 1922-   Kew, Eng., 1922-
Issued monthly. An abstracting service covering the more important recent literature dealing with the diseases of plants, except those caused by plant parasites. The major reference work for plant pathologists. World coverage. Abstracts by specialists. Has monthly author indexes and annual author and subject indexes.

B161 HEDWIGIA; organ für kryptogamenkunde und phytopathologie nebst repertorium für literatur. v. 1-82, 1852-1944. Dresden, Heinrich, 1852-1944. 82 v.
Bibliographical "Beihefte" appear, two to a volume, each in two parts: A, "Referate und kritische besprechungen" selectively abstracts world literature; B, "Neue literatur" is a bibliographical list only, more complete than A, listing items under such headings

as schizomycetes, myxomycetes, algae, fungi, etc.
B162 SOCIÉTÉ MYCOLOGIQUE de FRANCE. Bulletin trimestriel. Paris, 1885-
The "Revue bibliographique" section is an abstracting bibliography concerning fungi.
B163 ZEITSCHRIFT für pflanzenkrankheiten (pflanzenpathologie) und pflanzenschutz. v. 1-   1891-   Stuttgart, Ulmer, 1891-
Twelve issues a year. Publishes original papers and abstracts covering the world literature of plant pathology. Has annual author and subject indexes. Particularly useful for its coverage of European material not indexed in American reference works.
B164 ZENTRALBLATT für bakteriologie, parasitenkunde und infektionskrankheiten. Zweite abteilung: Allgemeine, landwirtschaftliche, technische, ... nahrungsmittel-bakteriologie und mykologie, protozoologie, pflanzenkrankheiten (einschliesslich). v. 1- 1895-   Jena, Fischer, 1895-
Issued irregularly. The "Referate" section abstracts material concerning plant diseases caused by organisms. Each volume has an author and subject index.

Bibliographies and Indexes

B165 AINSWORTH, G. C. The plant diseases of Great Britain; a bibliography. London, Chapman and Hall, 1937. 273 p.
"Attempts to collect together . . . the key references for the principal plant diseases of Great Britain so that descriptions of the symptoms, the causal agents, and methods of control may be quickly found and additional information discovered."
B166 ATANASOFF, D. Virus diseases of plants, a bibliography. Sofia, Houdijnik, 1934. 219 p.
----- Supplements. In Phytopathologische zeitschrift 10(1937):339-463; 12(1939/40):511-584.
B167 BARNES, H. V. and ALLEN, J. M. A bibliography of plant pathology in the tropics and in Latin America. (U.S.D.A. Bibliog. B. 14) 1951. 78 p.
Covers the years 1937-1949. See bibliographies by Cook and Otero (B176 and B177) for references on plant pathology in the tropics before 1937.
B168 BERLIN. BIOLOGISCHE ZENTRALANSTALT für LAND- und FORSTWIRTSCHAFT in BERLIN-DAHLEN. Bibliographie der pflanzenschutzliteratur. 1914/19-   Berlin, Parey, 1921-
A bibliography of world literature on plant protection and plant diseases. No abstracts. Author index to each volume. Arranged by broad subjects. Previously issued by Berlin (or Germany) Biologische Reichsanstalt für Land- und Forstwirtschaft. Volumes for 1940-1945 were edited by Dr. J. Bärner and published in 1953. Continues Jahresbericht über das Gebiet der Pflanzenkrankheiten (B172).
B169 COMMONWEALTH MYCOLOGICAL INSTITUTION. Bibliography of systematic mycology. (Mimeographed Publications no. 7-8.) Kew, Eng., 1948-1949.
B170 COOK, M. T. Host index of virus diseases of plants. In Journal of

Agriculture of the University of Puerto Rico 19(1935):315-406; Sup. 1, 20(1936):691-727; Sup. 2, 22(1938):411-435.

B171 COOK, M. T. Index of the vectors of virus diseases of plants. In The Journal of Agriculture of the University of Puerto Rico 19(1935):407-420; Sup. 1, 20(1936):729-739; Sup. 2, 22(1938): 437-439.

B172 JAHRESBERICHT über das gebiet der pflanzenkrankheiten ... v. 1-16, 1898-1913. Berlin, Parey, 1899-1917. 16 v.
Continued as Bibliographie der Pflanzenschutzliteratur (B168).

B173 KRIEGER, L. C. C. Catalogue of the mycological library of Howard A. Kelley. Baltimore, 1924. 260 p.
Lists by author an outstanding private library of between seven and eight thousand titles. Includes pamphlets, reprints, and important papers in journals.

B174 LINDAU, G. and SYDOW, P. Thesaurus litteraturae mycologicae et lichenologicae ratione habita praecipue omnium quae adhuc scripta sunt de mycologia applicata quem congresseunt. Lipsiis, Borntraeger, 1908-1917. 5 v. Reprint ed., New York, Johnson Reprint Corp., 1954.
V. 1 and 2 list alphabetically by author literature published to 1907. V. 3 lists works published from 1907 to 1910, including also some earlier items not listed in v. 1 and 2. V. 4 and 5 list under a classified subject arrangement the works noted in v. 1 to 3. An indispensable guide to the early literature of mycology.

B175 MYCOLOGIA ... v. 1-   January 1909-   Lancaster, Pa., 1909-
Bimonthly. Published for the New York Botanical Garden. Supersedes Journal of mycology. Includes "Index to American mycological literature," 1912 to date. Official organ of the Mycological Society of America, 1933 to date.

B176 OTERO, J. I. and COOK, M. T. A bibliography of mycology and phytopathology of Central and South America, Mexico, and the West Indies. In The Journal of Agriculture of the University of Puerto Rico 21(1937):249-286.

B177 OTERO, J. I. and COOK, M. T. Partial bibliography of virus diseases of plants. In The Journal of Agriculture of the University of Puerto Rico 18(1934):5-410; Sup. 1-3, 19(1935):129-313; 20(1936): 741-819; 22(1938):263-409.

B178 U.S. BUREAU of PLANT INDUSTRY. LIBRARY. Check list of publications of the state agricultural experiment stations on the subject of plant pathology, 1876-1920. (U.S.D.A. Libr. Bibliog. Contrib. 2) 1922. 179 p.

B179 U.S. BUREAU of PLANT INDUSTRY. LIBRARY. A check list of the publications of the Department of Agriculture on the subject of plant pathology, 1837-1918. (U.S.D.A. Libr. Bibliog. Contrib. 1) 1919. 38 p.

B180 WEISS, F. A. and O'BRIEN, M. J., comps. Index of plant diseases in the United States; issued by the Plant Disease Survey, Division of Mycology and Disease Survey, Bureau of Plant Industry, Soils, and Agricultural Engineering. pt. 1-5 (U.S. Bur. of Plant Ind., Soils, and Agr. Engin. Plant Disease Survey. Special P. 1.) 1950-1953.
The names and geographical locations of diseases are listed after the names of the species. The index section has an index to Latin

host names, families and genera, a common-name host index, and a list of authors of plant parasite names with recommended abbreviations.

Geography

B181 COMMONWEALTH MYCOLOGICAL INSTITUTE. Distribution maps of plant diseases. Map no. 1-   Kew, Eng., 1942-
Maps showing world distribution of various plant diseases are issued at intervals and are revised when necessary. Each map has a useful list of references printed on its verso.

Handbooks, Manuals, and Texts

B182 BAWDEN, F. C. Plant viruses and virus diseases. Ed. 3. Waltham, Mass., Chronica Botanica, 1950. 335 p.
An important text which reviews the literature of the subject and provides references at the end of each chapter.

B183 GRAM, E. and WEBER, A. Plant diseases in orchard, nursery, and garden crops, ed. by R. W. G. Dennis. New York, Philosophical Library, 1953. 618 p.
A well-illustrated reference manual with bibliographies at the end of each part.

B184 HOLMES, F. O. Handbook of phytopathogenic viruses. Minneapolis, Burgess, 1939.
References to literature and information concerning 129 viruses affecting seed plants. Bacteriophages treated in a supplement.

B185 SOCIETY of AMERICAN BACTERIOLOGISTS. Bergey's manual of determinative bacteriology, by Robert S. Breed, E. G. D. Murray, and A. Parker Hitchens assisted by 60 contributors. Ed. 6. Baltimore, Williams & Wilkins, 1948. 1,529 p.
A basic manual on bacteriology useful in plant pathology for information on bacteria and viruses causing plant diseases. Index of source and habitats has sections for plant diseases and plant hosts of viruses.

B186 SORAUER, P. Handbuch der pflanzenkrankheiten, begründet von Paul Sorauer, hrsg. von O. Appel. Berlin, Parey, 1928-   6 v.
In progress. The 6 volumes are being issued in parts and in various editions. A scholarly and thoroughly documented work started by one of the founders of the modern science of plant pathology. Has subject indexes and extensive bibliographies.

B187 STEVENSON, J. A. Foreign plant diseases . . . a manual of economic plant diseases which are new to or not widely distributed in the United States. Washington, 1926. 198 p.

B188 WESTCOTT, C. Plant disease handbook. New York, Van Nostrand, 1950. 746 p.
Written principally for the horticulturist. Information under name of disease and also under name of the garden plant. Has a bibliography.

B. Plant Sciences

Fungi

B189 AINSWORTH, G. C. and BISBY, G. R. A dictionary of the fungi. Ed. 3. Kew, Eng., Commonwealth Mycological Institute, 1950. 447 p.
Attempts to list all generic names of fungi. "There are in addition short accounts of the chief families, orders, and classes of fungi and of the bacteria and lichens; explanation of words used in mycology, etc." Appendix is G. W. Martin's "Key to the families of fungi."

B190 BESSEY, E. A. Morphology and taxonomy of fungi. Philadelphia, Blakiston, 1950. 791 p.
Bibliography, p. 660-756. Bibliographies at end of chapters.

B191 COMMONWEALTH MYCOLOGICAL INSTITUTE. Index of fungi. v. 1-   1940-   Kew, Eng., 1940-
Published as supplements to the Review of applied mycology (B160). Two numbers a year 1940-1947, issued annually 1948 to date. Continues Verzeichnis der neuen arten, varietäten, formen, namen und wichtigsten synonyme der pilze, 1932-1935 (Berlin, Borntraeger, 1944), and List of new species and varieties of fungi, 1936-1939 (Kew, Commonwealth Mycological Institute, 1950), both by Franz Petrak.

B192 FARLOW, W. G. Bibliographical index of North American fungi. Washington, Carnegie Institution, 1905.
V. 1, pt. 1, Abrothallus to Badhamia, all published. A subject and author catalog listing 653 titles. "A list of works on North American fungi, by W. G. Farlow and Wm. Trelease was issued in 1887 as Bibliographical Contribution 25 of the Library of Harvard University. A supplemental list of 756 titles was issued as Bibliographical Contribution 31 in 1888."

B193 FISCHER, G. W. The smut fungi, a guide to the literature, with bibliography. New York, Ronald Press, 1951. 387 p.
Has references on some 330 species of smut fungi and the diseases they cause. Represents information in more than 3,000 scientific papers, bulletins, and books.

B194 OUDEMANS, C. A. J. A. Enumeratio systematica fungorum. The Hague, Nijhoff, 1919-1924. 5 v.
Cites literature references through 1910 to fungi growing on European plants. Arranged according to host plants and their organs. Fungi are listed under the name of the organs. V. 5 is an alphabetical index. Table of abbreviations is in v. 1.

B195 SACCARDO, P. A. Sylloge fungorum omnium hucusque cognitorum. Patavii, 1882-1931. 25 v. in 27. Reprint ed., Ann Arbor, Mich., Edwards, 1944.
Contains descriptions in Latin of approximately 80,000 species of fungi. V. 16 and 18 have index to the preceding volumes. V. 17, contains a bibliography of mycological works. V. 19 and 20 contain the "Index iconum fungorum." A monumental work basic to the study of fungus diseases.

B196 SEYMOUR, A. B., comp. Host index of the fungi of North America. Cambridge, Mass., Harvard University Press, 1929. 732 p.
Gives for each host plant all fungi known to grow on it and for each fungus all hosts upon which it grows. Index includes 80,000 names. Arranged taxonomically.

B197 SNELL, W. H. Three thousand mycological terms. (R.I. Botanical
Club Publication 2.) Providence, R. I., 1936. 151 p.
"Includes technical terms ordinarily used in college courses in the morphology of the fungi and in mycology."
B198 WOLF, F. A. and F. T. The fungi. New York, Wiley, 1947. 2 v.
A text and reference work particularly valuable for its bibliographies. Has author and subject indexes.

FORESTRY AND FOREST PRODUCTS

The growth of forestry as a science and the increasing importance of forests to the world's economy has resulted in a large and interesting literature. It is estimated, for instance, that the Library of the U.S. Department of Agriculture has approximately 12,000 separate publications about trees and forests: over 400 books, pamphlets, and journal articles on the subject are received each month.

The reference works listed here cover both forestry and forestry products. They should also be supplemented by works noted elsewhere in this compendium. The Bibliography of agriculture (A24), for example, has monthly sections on forestry, forest management and finance, forest industries and trade, forest products, and wood technology. The Agricultural index (A23) lists forestry items under the appropriate subject headings. The Experiment station record (A22) had a monthly section on forestry until it ceased publication in 1946. Biological abstracts (A11) has a forestry unit, which does not, however, appear in each issue. Forestry abstracts are also published in Section D, "Abstracts of plant sciences," which is issued separately. Chemical abstracts (D8) has a section on cellulose and paper.

The worker in forestry literature will also need to consult the botanical and horticultural literature, and in particular such works as Rehder's The Bradley bibliography (B28) and Bibliography of cultivated trees and shrubs hardy in the cooler temperate regions of the Northern Hemisphere (B27).

Abstracting Journals

B199 APPLETON, WIS. INSTITUTE of PAPER CHEMISTRY. LIBRARY.
Bulletin. v. 1- September 1930- Appleton, Wis., 1930-
Issued monthly. Classified abstracting bibliography of periodical literature and patents in the field of paper manufacture and industry. Has sections on forests and forestry, straw, wood distillation, wood waste, etc. In addition to the abstracts, each issue contains a list of books and pamphlets and author index. Indexed annually by author and subject.
B200 COMMONWEALTH FORESTRY BUREAU. Forestry abstracts. v. 1-
1939- Oxford, 1939-
Issued quarterly. Formerly issued by Imperial Forestry Bureau. Systematically arranged bibliography of the world's literature. Has sections on general forestry, fundamental natural laws of the forest, silviculture, utilization and technology, injuries and protection, mensuration, management, economics, and policy. Annual author and subject indexes.
The Commonwealth Forestry Bureau also supplies a card index

56   B. Plant Sciences

service under the title Centralized title service. This service provides index cards which list all titles to appear in Forestry abstracts some five or six months before publication, plus all titles indexed but not published. There are three services: service A, comprising the whole of forestry, distributes 400 to 500 master cards a month; service B ("utilization" only) 150 to 200; and service C (the rest of forestry) 250 to 300. Citations are in English and give more bibliographical detail than the Abstracts. Each article is coded according to the Oxford system of decimal classification for forestry, and, where space allows, short abstracts are given.

B201   FORSTLICHE rundschau, ed. by Karl Abetz and Josef Köstler. v. 1-17, 1928-1945. Neudamm, Neumann, 1928-1945. 17 v.

-----   Beiheft. Internationale Titelsammlung für das jahr 1937-1938. Neudamm, Neumann [1938]-1939. 3 v.

Title varies. Vols. 1-16 received by U.S.D.A. Library. Note from publisher stated publication ceased at end of 1945. Abstracting journal covering the world literature of forestry and related subjects. Classified by subject. Indexed by author.

Bibliographies and Indexes

B202   BIBLIOGRAPHIA forestalis; forstliche bibliographie der Internationalen Forstzentrale. Bibliographia forstiere du Centre International de sylviculture. Hrsg. v. Josef Köstler. 1941-1942. Berlin, 1942-1943. 2 v.

Title bibliography of literature in various languages arranged according to the classification system used for the scientific sections of the International Forestry Centre. Titles in languages other than French, German, English, Italian, and Spanish are translated. Each volume contains more than 3,000 titles. Indexed by author. Preliminary material and section titles in several languages.

B203   A BIBLIOGRAPHY on woods of the world, exclusive of the temperate region of North America and with emphasis on tropical woods. Prep. by . . . G. P. Ahern and . . . H. K. Newton. (Tropical Plant Research Foundation, Scientific Contrib. no. 10; issued also as A.S.M.E. Research publication.) New York, American Society of Mechanical Engineers, 1928. 77 p.

Geographically arranged title bibliography with subject index. Based on a bibliography compiled by Samuel J. Record, entitled "Bibliography of the woods of the world with emphasis on tropical woods."

B204   BUNDESANSTALT für FORST- und HOLZWIRTSCHAFT, REINBEK. Bibliographie des ausländischen forst- und holzwirtschaftlicher schrifttums, 1948/49-   June 1950-

Issued annually.

B205   DAYTON, W. A. United States tree books, a bibliography of tree identification. (U.S.D.A. Bibliog. B. 20) 1952. 32 p.

Lists 381 titles published in the twentieth century.

B206   FOOD and AGRICULTURE ORGANIZATION of the UNITED NATIONS. Bibliography of forestry and forest products. September 1949- [Washington, 1949-   ]

Issued irregularly. "Reprinted by special premission from the Forestry and Forest products sections of the Bibliography of agriculture (A24)."

B207 GRÜNWOLDT, F. Die dokumentation in der forstwirtschaft. (Zeitschrift für Weltforstwirtschaft, v. 7. Beiheft.) Berlin, 1940. 87 p.
Contains three bibliographies: a list of international forestry bibliographies, a list of national forestry bibliographies, a list of special bibliographies of use in forestry.

B208 KRAEMER, J. H., comp. Wood conservation bibliography. (U.S. Off. Indus. and Com. Dom. Com. Ser. 30) 1952. 77 p.

B209 MUNNS, E. N., comp. A selected bibliography of North American forestry. (U.S.D.A. Misc. P. 364) 1940. 1,142 p.
A bibliography of the "more important literature" published in the United States, Canada, and Mexico before 1930. Classified, with author index.

B210 OXFORD UNIVERSITY. IMPERIAL FORESTRY INSTITUTE. Current monthly record of forestry literature. no. 1-43, January/February 1936-December 1939. Oxford, 1936-1939. 43 nos.
Issued monthly. Classified title bibliography. Not indexed. Succeeded by Forestry abstracts (B200).

B211 PARIS. INSTITUTION NATIONAL du BOIS. CENTRE de DOCUMENTATION. Bulletin bibliographique. 1- January 1950-
Issued monthly.

B212 U.S. FOREST SERVICE. FOREST PRODUCTS LABORATORY. List of publications on chemistry of wood and derived products. (R238) 1952. 41 p.

B213 U.S. FOREST SERVICE. FOREST PRODUCTS LABORATORY. List of publications on logging, milling, and utilization of timber products. (R790) 1953. 20 p.

B214 U.S. FOREST SERVICE. FOREST PRODUCTS LABORATORY. List of publications on mechanical properties and structural uses of wood and wood products. (R200) 1953. 51 p.

B215 U.S. FOREST SERVICE. FOREST PRODUCTS LABORATORY. List of publications on pulp and paper. (R444) 1953. 41 p.

B216 U.S. FOREST SERVICE. FOREST PRODUCTS LABORATORY. List of publications on the growth, structure, and identification of wood. (R177) 1952. 23 p.
Classified list in tabular form. Title, author, and publication date are given in parallel columns.

B217 U.S. FOREST SERVICE. FOREST PRODUCTS LABORATORY. List of publications on the seasoning of wood. (R446) 1953. 15 p.

B218 U.S. FOREST SERVICE. FOREST PRODUCTS LABORATORY. List of publications on wood finishing subjects. (R454) 1953. 16 p.

B219 U.S. FOREST SERVICE. FOREST PRODUCTS LABORATORY. List of publications on wood preservation. (R704) 1952. 29 p.

B220 U.S. FOREST SERVICE. FOREST PRODUCTS LABORATORY. List of publications relating to fungus defects in forest products and to decay in trees. (R508) 1952. 22.

B221 U.S. FOREST SERVICE. FOREST PRODUCTS LABORATORY. Some reference books on domestic and foreign woods. (R1479) 1945. 13 p.
List of books and pamphlets describing the characteristics, occurrence, and uses of various woods. References are grouped by

countries or regions and are in chronological order within each group, the most recent placed first. Prices are included.

B222 U.S. FOREST SERVICE. LIBRARY. Forestry current literature, February 1904-July/December 1933. v. 1-7, January 1934-April 1940. Washington, 1904-1940.
Title varies. May 1910-July 1919 printed in <u>American forestry</u>. March 1922-March 1927 in <u>Journal of forestry</u>. Superseded by U.S. Department of Agriculture Library, <u>Bibliography of agriculture</u> (A24). An index of periodicals and books on forestry and related subjects received in the Forest Service Library. Prepared for current use. Very few volumes indexed.

B223 U.S. FOREST SERVICE. LIBRARY. Publications on forestry, 1935-1940; an annotated list including reviews, abstracts, and other information . . . Washington, 1941. 2 v.

B224 U.S. NATIONAL ARCHIVES. Preliminary inventory of the records of the Forest Service, comp. by Harold T. Pinkett. (Preliminary Inventory, no. 18.) 1949. 17 p.

B225 WEST, C. J. Pulp and paper manufacture, bibliography and United States patents, 1931-   New York, Technical Association of the Pulp and Paper Industry, 1931-
Issued annually. Before 1931 published annually in the society's <u>Technical association papers</u> and in the <u>Paper trade journal</u>. Has also been cumulated in volumes covering the following periods: 1900-1928, 1928-1935, 1936-1945, 1946-1950. Cumulations do not list patents.

Dictionaries

See Section G, Addendum, for additional material.

B226 AMERICAN PAPER and PULP ASSOCIATION. Dictionary of paper, including pulps, boards, paper properties, and related papermaking terms. New York, 1940. 365 p.
Contains, in addition to the dictionary proper, sections on the classification and definitions of pulp and the classification of waste materials used in the paper and board industries.

B227 ARO, P., and others, eds. Suomalais, ruotsalais, saksalais, englantilainen metsäsanakirja ... Finnish, Swedish, German, English forest dictionary. Helsinki, Kustannusosakeyhtiö Otara, 1944. Various paging.
Preliminary material and contents in Finnish, Swedish, German, and English. List of Finnish words with equivalents in the other three languages. Indexed by Swedish, German, and English words. Index of Latin names is also included.

B228 L'ASSOCIATION des INGENIEURS FORESTIERS de la PROVINCE de QUEBEC. Vocabulaire forestier. Quebec, 1946. 502 p.

B229 BOULTON, E. H. B. A dictionary of wood. London, Nelson, 1938. 205 p.
Illustrated and arranged alphabetically by kind of wood. Gives for each genus: species, family, distribution, general properties, size and availability, uses, and finishes.

B230 BRITISH STANDARDS INSTITUTION. British standard specifications. Nomenclature of commercial timbers, including botanical names

and sources of supply. London, 1946. 101 p.
In tabular form. Standard name, botanical species, sources of supply, remarks, and other commercial or botanical names are shown in parallel columns. Indexed and contains a list of the authors, with dates of birth and death, of the botanical names used.

B231 BRUTTINI, A. Dictionnaire de sylviculture en cinq langues. Paris, Lechevalier, 1930. 384 p.
Defines 4,600 terms related to temperate and tropical forests.

B232 BUCHHOLZ, E. Kleines russisch-deutsches forstwörterbuch. Berlin, Ost-Europa-Verlag, 1943. 138 p.
Russian-German forestry dictionary containing about 6,000 terms. Includes terms on botany, biology, zoology, soil science, and hunting.

B233 CORKHILL, T. A glossary of wood. London, Nema Press, 1948. 655 p.
Illustrated and contains about 10,000 terms pertaining to timber and its uses and the wood-working trades. Appeared serially in the monthly publication Wood.

B234 DEEN, J. L., BENSON, A. B, and DANNFELT, J. J. A Swedish-English vocabulary for foresters. (Yale University School of Forestry B. 40) 1935. 83 p.

B235 GARCÍA-PIQUERA, C. Glosario de terminologia forestal. (Puerto Rico. Departamento de Agricultura y Comercio. Monografia 5) 1955. 172 p.
An English-Spanish, Spanish-English glossary of forestry terms. Has bibliography of items relating to South American forestry and nomenclature.

B236 LITSCHAUER, R. von. Vocabularium polyglottum vitae silvarum. Waldbiologisches fachwörterbuch auf der grundlage der wissenschaftlichen nomenklatur. Lateinisch-Deutsch-Französisch-Englisch-Spanisch-Russisch. Hamburg and Berlin, Parey, 1955. 126 p.
Provides English, French, German, Spanish, and Russian equivalents of Latin terms for plants, animals, diseases, harmful insects, and other biological words and phrases connected with forestry.

B237 MEYER, H. Buch der holznamen. The book of wood names. Les noms de bois. Hannover, Schaper, 1936. 564 p.
In four languages: German, English, French, and Spanish. Arranged alphabetically by common or catch-word name. Gives species, family, and origin.

B238 MRUGOWSKI, H. Fachworterbuch fur die holzwirtschaft. Hannover, Schaper, 1948. 2 v.

B239 RABER, O., comp. German-English dictionary for foresters. Washington, U.S. Forest Service, 1939. 346 p.
Loose-leaf. Includes terms encountered in the botanic, entomological, geological, soil, chemical, and engineering aspects of forestry as well as terms in the fields of recreation, game management, soil conservation, land use, and forest economics.

B240 SHEPARD, W., comp. Spanish terms for forest rangers. n. p., 1921. 44 p.
Compilation of "colloquial words and terms most frequently met with in the conduct of Forest Service business among the Spanish-speaking people of the Southwest." Terms are arranged under the

following headings: General administration, Forest management, Grazing, Lands, Fire, Law, Surveying and mapping, Travel and subsistence, Accounts, Weights and measures. Indexed by English terms.

B241 SOCIETY of AMERICAN FORESTERS. COMMITTEE on FORESTRY TERMINOLOGY. Forestry terminology, a glossary of technical terms used in forestry. Rev. ed. Washington, 1950. 93 p.
"Limited to (1) terms used in a special sense by foresters, and (2) terms from other sciences and industries the meaning of which a forester should know and which may not be defined in other glossaries or texts likely to be easily available to the average forester." List of glossaries in allied sciences, p. 2.

B242 VOGT, W. Holzfachwörterbuch. Berlin-Tempelhof, The Author, 1933. 2 v.

Directories

See note on directories, p. 15.

B243 COMMONWEALTH FORESTRY BUREAU. The chief research centres for forestry and wood utilization in continental Europe, excluding most of the Balkan Peninsula and Eastern Europe. In Forestry Abs. 13(1951):119-131.

B244 EMPIRE FORESTRY ASSOCIATION. The empire forestry handbook, 1952. London, 1953. 210, 90 p.
Names and addresses of members of the Empire Forestry Association, list of forest services and forest research and educational institutions of the British Empire, list of principal forest research organizations of the world, forestry societies of the British Empire, and a list of forestry periodicals. A section, "Forest resources of the world," by Sir Hugh Watson, is also included. Standard nomenclature of timbers on the British market is given in an appendix.

B245 FOOD and AGRICULTURE ORGANIZATION of the UNITED NATIONS. Directory of forestry schools. (F.A.O. Forestry and Forest Product Studies, no. 10) New York, Columbia University Press, 1954. 305 p.
Arranged by country. Has information on curricula, degrees, students, faculties, facilities for training, and living costs.

B246 GILL, T. and DOWLING, E. C., comps. The forestry directory. Washington, American Tree Association, 1949. 420 p.
Reference volume on forestry activities in the United States and Canada. Includes accounts of forestry in the U.S. Department of Agriculture and other government agencies. Lists national, regional, and trade organizations and forestry schools and contains a section on forest miscellany. Subject index.

B247 GRÜNWOLDT, F. Internationales adressbuch für forstwirtschaft, holzwirtschaft. Jagd und naturschutz. International index to forestry and allied organizations . . . pt. 1. Neudamm, Neumann, 1938. 94 p.
Pt. 1, International and North America. Directory listing the names and addresses of organizations, associations, committees,

Forestry and Forest Products 61

federal and state departments, universities, etc. Includes maps and charts. Indexed. In German, English, and French.

B248 LOCKWOOD'S directory of the paper and allied trades. Ed 9, 1954. New York, Lockwood Trade Journal, 1953. 1,600 p.

Issued annually. Partial contents: Classified lists of paper mill and converted products of the United States and Canada; Classified lists of pulp mill products, Canada and the United States; Coated paper manufactures; Converters of paper and paper board in Canada and the United States; Envelope manufacturers; Exporters and importers of paper; Glazed and coated paper manufacturers; List of mill officials; Paper and pulp mills of Canada, Cuba, Mexico, South America, and the United States; Paper bag and box manufacturers; Paper merchants, United States and Canada; Prepared roofing manufacturers; Trade associations; Watermarks and brands; Wood pulp agents, importers, and exporters.

B249 LUMBERMAN'S handbook and directory of the Southern forest industries, 1952. Seattle, Wash., The Lumberman, 1951. 252 p.

Includes a section entitled, "Sawmills, logging concentration yards, and re-manufacturing plants" for states in the region, by cities.

B250 LUMBERMAN'S handbook and directory of the Western forest industries, 1952. Seattle, Wash., The Lumberman, 1951. 381 p.

Includes a section on sawmills and logging camps for states in the region, by cities.

B251 PAPER and pulp mill catalogue, 1949-1950. Ed. 26. Chicago, Fritz Publications, 1949. 520 p.

Issued annually. A directory of manufacturers of supplies and equipment for paper and pulp mills. Trade names index and a directory of institutions of the United States and Canada offering courses in pulp and paper manufacture are also included. Tables, definitions, and other information for pulp and paper.

B252 THE PAPER yearbook, 1954. Chicago, Davidson, 1954. 570 p.

Issued annually. Directory of paper and paper products giving information regarding construction, sizes, uses, etc., of the items and list of manufacturers. Various tables, charts, paper testing procedures, and the story of paper making are also included.

B253 PHILLIPS' paper trade directory of the world, 1954. London, 1954. 806 p.

Issued annually. Classified directory of paper mills, manufacturers, etc., in Great Britain and foreign countries. Lists of watermarks and trade names and British paper, wood pulp, and allied trades associations are included. Glossary of trade terms in English, French, German, Spanish, Italian, and Swedish, p. x-xxix.

B254 POST'S paper mill directory. 1954 ed. New York, Paper Mill News, 1954. 672 p.

Issued annually. Lists, arranged by states and cities and classified by product, of pulp and paper mills of the United States and Canada. Also included are lists of mill officials and associations of and affiliated with the pulp and paper industry.

B255 TIMBERMAN. Forest products industry directory of Western North America, 1950. Portland, Ore., 1949. 455 p.

Lists, arranged alphabetically by states and cities, of sawmills and logging companies and manufacturers of wood products with names of officers and data on equipment and operations. Also includes lists of lumber wholesalers and exporters, trade associations, and forestry organizations. Company and individual name indexes.

B256 U.S. FOREST SERVICE. Organizational directory . . . March 1953. Washington, 1953. 122 p.

Issued annually. Lists names and headquarters of the Forest Service personnel responsible for various units, activities, and lines of work. Arranged according to regions, stations, etc. Includes individual name index.

B257 U.S. FOREST SERVICE. DIVISION of FOREST PRODUCTS. Lumber manufacturing, wood using, and allied associations. Washington, 1946. 14 p.

Classified directory. Gives for each association, name, address, principal officers, and date of latest grading rules for lumber or wooden products.

Trees - Lists and Descriptive Guides

B258 BRITTON, N. L. and SHAFER, J. A. North American trees: being descriptions and illustrations of the trees growing independently of cultivation in North America, north of Mexico, and the West Indies. New York, Holt, 1908. 894 p.

Manual arranged according to families. Includes a glossary and English and Latin name indexes.

B259 COLLINGWOOD, G. H. Knowing your trees. With 529 photographs showing typical trees and their leaves, bark, flowers, and fruits. 5th and rev. printing. Washington, American Forestry Association, 1941. 213 p.

Contains 101 tree descriptions.

B260 HOUGH, R. B. Handbook of the trees of the northern states and Canada east of the Rocky Mountains. Photo-descriptive. New York, Macmillan, 1947. 470 p.

Illustrations show trunk, branchlet, leaves, fruit, pods, seeds, etc. Maps show distribution.

B261 LITTLE, E. L., Jr. Check list of native and naturalized trees of the United States (including Alaska). (U.S.D.A. Agr. Handb. 41) 1953. 472 p.

Supersedes Misc. Circ. 92.

B262 REHDER, A. Manual of cultivated trees and shrubs hardy in North America, exclusive of the subtropical and warmer temperate regions. Ed. 2, rev. and enl. New York, Macmillan, 1940. 996 p.

Systematically arranged brief descriptions of over 2,550 species. Indexed by both Latin and common name. Glossary of botanical terms, p. 903-908.

B263 SARGENT, C. S. Manual of the trees of North America (exclusive of Mexico) with seven hundred and eighty-three illustrations by Charles Edward Faxon and Mary W. Gill. Ed. 2. Boston, Houghton, Mifflin, 1933. 910 p.

Systematically arranged. Describes each tree and gives information regarding its distribution. Includes a map showing the 8 regions into which the country is divided according to the prevailing character of the trees. Indexed by both Latin and common name.

B264 SARGENT, C. S. The silva of North America; a description of the trees which grow naturally in North America exclusive of Mexico . . . il. with figures and analyses drawn from nature by Charles Edward Faxon. Boston, Houghton Mifflin, 1892-1902. 14 v.
   Botanical descriptions with excellent illustrations and bibliographical footnotes. Mentions uses and purposes for which cultivated. General index to all volumes is contained in v. 14.
B265 SCHMUCKER, T. The tree species of the northern temperate zone and their distribution. (Silvae Orbis 4) Berlin, Centre International de Sylviculture, 1942. 156 p. text; 250 p. maps.
   Text in French, German, English, Spanish, and Italian.
B266 U.S. FOREST SERVICE. Check list of the native and naturalized trees of the United States including Alaska. Washington, 1944. 325 p.
   Covers 79 families, 225 genera, 1,015 species, and 167 varieties. Arranged alphabetically by scientific names. Common name, references, and range are also given. Index of common names is included.

Handbooks and Manuals

B267 HOWARD, A. L. A manual of timbers of the world. Ed. 3. London, Macmillan, 1948. 751 p.
   Briefly describes important timbers and their uses. Has indexes to scientific and vernacular names and a list classifying timbers by country of origin.
B268 NATIONAL LUMBER MANUFACTURERS' ASSOCIATION. Lumber industry facts. Washington, 1950. 48 p.
   Statistical information, usually from U.S. government sources, concerning the lumber industry in the United States.
B269 RECORD, S. J. and HESS, R. W. Timbers of the New World. New Haven, Conn., Yale University Press, 1943. 640 p.
   Describes over 1,000 genera of trees in the Americas and evaluates their economic importance. Supplies basis for identification of trees and wood, and correlates the vernacular, trade, and scientific names. Extensive bibliography, p. 573-588.
B270 SOCIETY of AMERICAN FORESTERS. Forestry handbook, ed. by R. D. Forbes and A. B. Meyer. New York, Ronald, 1955. 23 sections.
   A key reference tool for foresters. Tabulates and organizes forestry knowledge. Provides information on topics such as log and yield measurement, fire protection, and range management. Defines terms, describes and illustrates tools and equipment. Each of the 23 sections has a selected bibliography.
B271 U.S. DEPARTMENT of AGRICULTURE. Trees. (Ybk. of Agr., 1949) 1949. 944 p.
   Signed articles on the importance of trees and forests in the national economy. Sections discuss famous trees, shade trees, small woodlands, wind breaks, Christmas trees, company and national forests, naval stores, insects, parasites and diseases, forest vacations, forests and wildlife, wood utilization, history of forestry, and forest resources. Vacation guide and glossary of

terms used by woodsmen are also included. Illustrated and indexed.

B272 U.S. FOREST SERVICE. FOREST PRODUCTS LABORATORY. Wood handbook, basic information on wood as a material of construction with data for its use in design and specification. (Slightly rev., June 1940) Washington, 1940. 326 p.

Glossary, p. 3-9; Standard lumber abbreviations, p. 95-97. Includes bibliographies.

B273 U.S. BUREAU of FOREIGN and DOMESTIC COMMERCE. BUSINESS INFORMATION SERVICE. American lumber industry, basic information sources. Washington, 1951. 12 p.

Cites bibliographies, statistical compilations, and other sources of information for the lumber industry.

Geography

B274 VAN ROYEN, W. Atlas of the world's resources: v. III, Forest and fishery resources. New York, Prentice Hall. [In preparation, 1957.]

See A107 for v. I.

Periodical Lists

B275 FOOD and AGRICULTURE ORGANIZATION of the UNITED NATIONS. Forestry abstracts coverage list, periodicals, and serials. (F.A.O. Forestry and Forest Product Studies, no. 7.) Rome, 1953. 180 p.

Prepared by the Commonwealth Forestry Bureau from its card index of the literature forming the basis of its quarterly publication, Forestry abstracts (B200).

B276 GRÜNWOLDT, F. Répertoire international des périodiques forestiers, sylviculture, économie du bois, protection de la nature et chasse d'après leur état au 1er janvier 1940. (Silvae Orbis 1) Berlin-Wannsee, Centre International de Sylviculture, 1940. 205 p.

Preliminary matter and half-titles in French, German, English, Spanish, and Italian. Directory of forestry serials arranged alphabetically by title under continents, countries, colonies, etc. For the different periodicals information is given as follows: full title, abbrevitation of title, issuing agency, editorial office, editor, date of founding, and price. Has alphabetical, geographical, and subject indexes.

B277 U.S. FOREST SERVICE. LIBRARY. A union check list of forestry serials, comp. by ... Helen Moore. Washington, 1936. 146 p.

List of approximately 1,800 American and foreign periodicals, reports, etc., in libraries of the United States and Canada.

Statistics

B278 COMITÉ INTERNATIONAL du BOIS. Year-book of world timber trade, 1933/1932-   Vienna, 1934-

Has probably ceased publication. V. 5, 1937/1936(1938) was last vol. received at U.S.D.A. Library. Statistical tables showing exports, imports, and value of timber by classes and by country and world trade in sawn and planed softwood, pulpwood, and plywood.

B279 FOOD and AGRICULTURE ORGANIZATION of the UNITED NATIONS. Timber statistics for the years 1946-1947. Geneva, 1948. 39 p.
In French and English. Collection of statistical tables showing world output, stocks, purchase and sales, exports and imports of sawn soft and hardwood, pitprops, buildingboards, and plywood by country. See (B280) for statistics of later years.

B280 FOOD and AGRICULTURE ORGANIZATION of the UNITED NATIONS. Yearbook of forest product statistics. Annuaire statistique des produits forestiers, 1947-   Washington and Rome, 1948-
Issued annually. In French and English; Spanish supplement. Statistical tables showing world production and exports and imports of wood, processed wood, wood pulp and pulp products, and forest products other than wood.

B281 INTERNATIONAL INSTITUTE of AGRICULTURE. International trade of wood, statistical figures for the years 1925 to 1939. Rome, 1944. 597 p.
In French and English. Tables arranged alphabetically by country under continents show exports, imports, and value of wood and wood products. Trade in wood pulp is shown in an appendix. Indexed by country.

B282 INTERNATIONAL INSTITUTE of AGRICULTURE. International yearbook of forestry statistics. Rome, 1933-1943. 5 v.
In French and English. Collection of statistics on forest area, wood production, and trade in wood and wood products. First edition contains statistics relating to 1932 and earlier years. Second edition was issued in 4 v.: v. 1, 1933-1935 (Europe and U.S.S.R.); v. 2, 1933-1935 (American); v. 3, Africa (contains trade statistics for the period 1930-1939); v. 3, sup., Africa (contains additional material and indexes). Earlier forestry statistics are in International yearbook of agricultural statistics (F17), 1929-1930 and 1930-1931.

Miscellaneous

B283 FORD ROBERTSON, F. C. The Oxford system of decimal classification for forestry, authorized English version. Farnham Royal, Commonwealth Agricultural Bureaux, 1954. 115 p.
Contains the Oxford system of decimal classification for forestry as approved by the International Union of Forest Research Organizations and the Food and Agriculture Organization. Has a history of the International classification of forest literature. There are separate alphabetical indexes of subject and form numbers, and for geographical numbers.

B284 MEELIG, M. L., comp. Subject headings used in the Library of New York State College of Forestry. Syracuse, N.Y., 1949. 128 p.

B285 U.S. FOREST SERVICE. LIBRARY. Subject headings for forestry libraries. Washington, 1941. 178 p.
Revision of "List of subject headings used in the catalog of the Forest Service Library," issued in 1934.

SECTION C
ANIMAL SCIENCES

The subject fields considered in Section C are: economic zoology in general, animal (livestock) husbandry, poultry husbandry, veterinary medicine, economic entomology, apiculture, pest control, and commercial fishing and fisheries. A brief survey of the literature of wildlife management, defined broadly to mean chiefly undomesticated birds and mammals, is included in the following section on economic zoology.

ECONOMIC ZOOLOGY

The term "economic zoology" includes those areas of study and research which attempt to control animal life in the economic, sanitary, or medical interests of man. The applied, or economic, emphasis distinguishes this field from pure, or fundamental, zoology, which is interested primarily in classifying and describing animal life. A species has economic significance when it is established as an agent of destruction or as a positive economic product in its own right. The facts determined by the zoologist then become of value to the applied scientist, who attempts to develop controls over the species for economic reasons.

This distinction, applying both to the field of general zoology and to its specialized subdivisions, has been used in selecting the literature cited in the following sections. As much as possible, discussion is limited to the literature of the economic or applied phases of these sciences. Since the distinction cannot be maintained absolutely, however, the literature described frequently pertains to a fundamental science. This is particularly true where effective reference information sources have not yet been established for an applied science, or where the line between fundamental and applied sciences is difficult to define.

Bibliographies of Bibliographies

The following detailed guides to the literature of the general zoological sciences are recommended. Additional guides to zoological literature, such as Bourlière's Éléments d'un guide bibliographique du naturaliste, Wood's Introduction to the literature of vertebrate zoology, etc., are described in Winchell (A8).

Economic Zoology 67

C1   SMITH, R. C. Guide to the literature of the zoological sciences. Ed. 3. Minneapolis, Burgess, 1952. 133 p.
      Written by a professional entomologist, this guide is a practical working tool for the zoological specialist. There is an annotated list of journals, information on library usage, instructions on preparation of bibliographies, etc.
C2   STEVENS, F. C. Books of reference in zoology, chiefly bibliographical. In Soc. Bibliog. Nat. Hist. J. 3(1955):72-91.
      Includes a list of general reference books and reference books pertaining specifically to the various subdivisions of the animal kingdom. Includes many old and obscure works.

Abstracting Journals

Biological abstracts (A11) attempts to include as much of the research literature of zoology as possible, including that from economic and applied fields. It is the most widely used abstracting service for the life sciences in English-speaking countries. Germany has produced a number of zoological abstracting and indexing journals: Archiv für Naturgeschichte (1835-1926), Zoologischer Jahresbericht (1879-1913), Zoologischer Bericht (1922-1943/44), etc. Resumptio genetica (B130) indexes, and abstracts selectively, the genetic literature of zoology.

C3   COMMONWEALTH BUREAU of AGRICULTURAL PARASITOLOGY (HELMINTHOLOGY). Helminthological abstracts. v. 1- 1932-   Farnham Royal, Bucks, Eng., 1932-
      Five or six issues annually. The last issue is published late and contains author and subject indexes, abstracts of literature for the year omitted from previous issues, a list of journals covered, etc. Preceded by the Bibliography of helminthology (1930-1933), which was incorporated in Helminthological abstracts beginning with 1934. A specialized abstracting service of the literature on worms that are parasitic in animals, and therefore of interest to animal scientists.
C4   U.S. FISH and WILDLIFE SERVICE. Wildlife review. v. 1-   1935- Washington, 1935-
      Published somewhat irregularly, usually four issues a year. Abstracts chiefly North American literature, mostly publications on game, fur, and pest vertebrates. Excludes most publications on taxonomy, anatomy, experimental biology, popular articles and books, etc.
C5   U.S. FISH and WILDLIFE SERVICE. Wildlife abstracts, 1935-51. An annotated bibliography of the publications abstracted in the Wildlife review, nos. 1-66. Washington, 1954. 435 p.
      More than 10,000 abstracts cited, with author and subject index and list of journals in which original articles appeared.

Bibliographies and Indexes - General

Only a few of the bibliographies on zoological subjects can be cited here. For extensive lists, see the Bibliographic index (A2), Theodore

C. Animal Sciences

Besterman's World bibliography of bibliographies (A1), Smith (C1), etc.
Some literature on the breeding and general economic management of animals other than common livestock, is indexed in the Bibliography of agriculture (A24) and abstracted in Animal breeding abstracts (C64). See also the Index-catalogue of medical and veterinary zoology (C237).

C6 BIBLIOGRAPHIA zoologica. v. 1-43, 1896-1934. Zurich, Sumptibus Councilii Bibliographici, 1896-1934.
C7 BRITISH MUSEUM (NAT. HIST.). LIBRARY. Catalogue of the books, manuscripts, maps, and drawings in the British Museum (Natural History). London, 1903-1940. 8 v.
An author catalog.
C8 INTERNATIONAL catalogue of scientific literature ... 1901-1914. London, Pub. for the International Council by the Royal Society of London, 1902-1921. 254 v.
Section N - Author and subject index to zoological literature, merged in 1906 with the Zoological record (C10).
C9 U.S. ARMY MEDICAL LIBRARY. Index-catalogue of the library of the Surgeon General's Office, United States Army ... authors and subjects. Ser. 1-4, v. 10. Washington, 1880-1948.
The subsequent U.S. Armed Forces Medical Library, Catalog (1948-   ), the Current list of medical literature, and the Quarterly cumulative index medicus, contain references to a great deal of literature of interest to animal scientists.
C10 ZOOLOGICAL record. v. 1-   1864-   London, Zoological Society of London, 1865-
The most complete indexing service in zoology, particularly for systematic literature.

Bibliographies and Indexes - Early

Bibliographies of eighteenth- and early nineteenth-century literature on applied animal sciences are rare, and the distinction between fundamental and applied sciences is not clearly maintained. Significant retrospective bibliographies that are particularly effective in listing this earlier literature are given here.

C11 AGASSIZ, L. Bibliographia zoologiae et geologiae. A general catalogue of all books, tracts, and memoirs on zoology and geology. Cor., enl., and ed. by H. E. Strickland. London, Ray Society, 1848-1854. 4 v.
C12 BIBLIOTHECA zoologica. I. Verzeichniss der schriften über zoologie welche in den periodischen werken enthalten und vom jahre 1846-1860 selbständig erschienen sind. Leipzig, Engelmann, 1861. 2 v.
C13 BIBLIOTHECA zoologica. II. Verzeichniss der schriften über zoologie welche in den periodischen werken enthalten und vom jahre 1861-1880 selbständig erschienen sind. Leipzig, Engelmann, 1887-1923. 8 v.
C14 ENGELMANN, W. Bibliotheca historico-naturalis. Verzeichnis der bücher über naturgeschichte, welche in Deutschland, Scandinavien,

Holland, England, Frankreich, Italien, und Spanien 1700-1846 erschienen sind ... Mit einem namen- und sach-register. Leipzig, The Author, 1846. 786 p.
Includes zoological literature.

C15 REUSS, J. D. Repertorium commentationum a societatibus litterariis editarum. Gottingae, Dieterich, 1801-1821. 16 v.
V. 1 covers the zoological sciences and indexes publications of learned societies issued up to 1800.

C16 ROYAL SOCIETY of LONDON. Catalogue of scientific papers, 1800-1900. London, Clay, 1867-1902. 19 v.
An author index to literature of zoology and other sciences. No subject index to zoology sections available.

Bibliographies and Indexes - Special

C17 ARCTIC INSTITUTE of NORTH AMERICA. Arctic bibliography. Prep. for and in cooperation with the Department of Defense under the direction of the Arctic Institute of North America. V. 1-   1953-   Washington, Department of Defense, 1953-
V. 1-3, basic bibliography; v. 4- , annual supplements with self-contained subject indexes. First six volumes include 38,410 citations, mostly with abstracts and contents notes.
A recently initiated bibliographic series covering the literature dealing with the Arctic, including that concerned with zoological life forms.

C18 COUES, E. American ornithological bibliography. Washington, 1879-1880. 4 v. in 1.

C19 CUTHBERTSON, S. A preliminary bibliography on the American fur trade. St. Louis, U.S. Department of the Interior, National Park Service, Jefferson National Expansion Memorial, 1939. 191 p.

C20 DUMONT, P. A. National wildlife refuge bibliography. U.S. Fish and Wildlife Serv. (Wildlife L. 334) 1951. 53 p.

C21 DUTILLY, A. A. A bibliography of reindeer, caribou, and musk-ox. U.S. Quartermaster Corps. Mil. Planning Div. (Rpt. 129) 1949. 462 p.

C22 HICKEY, J. J. and WAGNER, F. H. A check list of technical game bulletins published by the state game departments. In Wildlife Review 72(March, 1953):1-43.
Covers 1938-1951. Includes a list of state game department periodicals.

C23 IMPERIAL BUREAU of ANIMAL BREEDING and GENETICS. Bibliography on fur breeding. Edinburgh, 1931. 37 p.

C24 IRWIN, R. British bird books; an index to British ornithology, A.D. 1481 to A.D. 1948. London, Grafton, 1951. 398 p.
Cites bibliographies and periodicals useful to British ornithologists, as well as books and journal articles.

C25 JOHANSSON, I. Förteckning över pälsdjurslitteratur. Malmö, Sydsvenska Dagbladets Aktiebolag, 1945. 50 p.
A bibliography on fur-bearing animals.

C26 PHILLIPS, J. C. American game mammals and birds; a catalogue of books, 1582 to 1925, sport, natural history, and conservation. Boston, Houghton Mifflin, 1930. 638 p.

C27 WEED, C. M. A partial bibliography of the economic relations of North American birds. In his Birds in their relations to man. Ed. 4. Philadelphia, Lippincott, 1935. p. 355-407.

C28 YEAGER, L. E. A contribution toward a bibliography on North American fur animals. (Ill. Nat. Hist. Survey. Biol. Notes, no. 16.) Urbana, Ill., 1941. 209 p.
Covers history, conservation, management, fur farming, with a list of journals.

Dictionaries

There are specialized dictionaries and terminological guides in many of the fields of zoology, but no general dictionary of zoology in the English language. Many of the general science dictionaries, and some of the medical dictionaries, include zoological terms. These will be found described in Winchell (A8). There are a number of informative books on terminology and some foreign dictionaries intended for zoologists.

C29 AMERICAN ASSOCIATION for the ADVANCEMENT of SCIENCE. Zoological names. A list of phyla, classes, and orders. Durham, N. C., 1949. 24 p.

C30 ARTSCHWAGER, E. F. Dictionary of biological equivalents, German-English. Baltimore, Williams & Wilkins, 1930. 239 p.

C31 BLANCHARD, R. A. É. Glossaire allemand-francais des terms d'anatomie et de zoologie. Paris, Asselin et Houzeau, 1908. 298 p.

C32 CARPENTER, J. R. An ecological glossary. Norman, University of Oklahoma Press, 1938. 306 p.

C33 HENDERSON, I. F. Dictionary of scientific terms; pronunciation, derivation, and definition of terms in biology, botany, zoology, anatomy, cytology, genetics, embryology, physiology. Ed. 5. Edinburgh, Oliver, 1953. 506 p.

C34 HIRSCH-SCHWEIGGER, E. Zoologisches wörterbuch. Berlin, De Gruyter, 1925. 628 p.

C35 JAEGER, E. C. A dictionary of Greek and Latin combining forms used in zoological names. Springfield, Ill., Thomas, 1930. 101 p.

C36 JAEGER, E. C. A source-book of biological names and terms. Springfield, Ill., Thomas, 1944. 256 p.

C37 MELANDER, A. L. Source book of biological terms. New York, Department of Biology, The City College of the City of New York, 1937. 157 p.

C38 NEAVE, S. A. Nomenclator zoologicus; a list of the names of genera and subgenera in zoology from the tenth edition of Linnaeus 1758 to the end of 1945. London, Zoological Society of London, 1939-1950. 5 v.

C39 NIEMANN, G. Zoologisches wörterbuch. Osterwieck-Harz, Zickfeldt, 1919. 221 p.

C40 WALLMEYER, B. Pelztragende tiere. Fur-bearing animals. Handbuch für die rauchwaren-wirtschaft ... deutsch-englische ausgabe. Frankfurt am Main, Schauer, 1951. 215 p.

C41 WOODS, R. S. Addenda to the Naturalist's lexicon. Comprehensive classified English-classical key to descriptive terms. Additions and emendations to the classical-English lexicon. Pasadena,

Calif., Abbey Garden Press, 1947. 47 p.
C42 WOODS, R. S. The naturalist's lexicon, a list of classical Greek and Latin words used, or suitable for use, in biological nomenclature, with abridged English-classical supplement. Pasadena, Calif., Abbey Garden Press, 1944. 282 p.
C43 ZIEGLER, H. E. Zoologisches wörterbuch; erklärung der zoologischen fachausdrücke. 3 Aufl. Jena, Fischer, 1927. 786 p.

Nomenclatural Guides

Nomenclatural guides for zoology are discussed in Smith (C1). The Zoological record (C10) lists new additions to zoological nomenclature, as well as literature published on changes. The changes are also listed and discussed in the various series published by the International Commission on Zoological Nomenclature, such as the Opinions and declarations (1943- date), and the Bulletin of zoological nomenclature (1943-date).

C44 CABLE, M. Introduction to ornithological nomenclature. Los Angeles, American Book Institute, 1955. 99 p.
C45 MAYR, E. Methods and principles of systematic zoology. New York, McGraw-Hill, 1953. 328 p.
C46 SCHENK, E. T. Procedure in taxonomy, including a reprint in translation of the Règles internationales de la nomenclature zoologique (International code of zoological nomenclature) with titles of and notes on the opinions rendered to the present date (1907-1947); completely indexed. Stanford, Calif., Stanford University Press, 1948. 93 p.

Directories

V. 2 of American men of science (A83), includes zoologists.

C47 CONSERVATION yearbook. 1952-   Washington, 1952-
Issued annually. Includes particularly complete directories of national, state, and private organizations and officials who are concerned with the conservation of national resources, including wildlife.
C48 INTERNATIONAL UNION of BIOLOGICAL SCIENCES. Index des zoologistes. (Sér. C. P. Diverses no. 5) Paris, 1953. 429 p. A worldwide directory.
C49 NATIONAL WILDLIFE FEDERATION. Directory of organizations and officials concerned with the protection of wildlife and other natural resources. Washington, 1955. 62 p.
Lists North American officials and organizations and Central and South American organizations. A similar directory, with varying titles, has been published since 1900 by various federal agencies, such as the Bureau of Biological Survey, the Fish and Wildlife Service, etc.
C50 ZOOLOGISCHES adressbuch. Namen und adressen der lebenden zoologen, anatomen, physiologen und zoopalaeontologen, sowie der künstlerischen und technischen hülfskräfte. Berlin, Friedlander,

1895-1901. 2 v.
A second edition, of 1109 pages, was published in 1911. A copy was not available for examination by the compilers.

Geography

C51 BARTHOLOMEW, J. G. Atlas of zoogeography. (Bartholomew's Physical atlas, v. 5) Edinburgh, 1911. 67 p.
About 200 maps, showing distribution of families and many genera and species of mammals, birds, reptiles, amphibians, fishes, insects, etc. Includes descriptive text and extensive bibliography.

C52 BEAUFORT, L. F. de. Zoogeography of the land and inland waters. London, Sidgwick & Jackson, 1951. 208 p.

C53 HESSE, R. Ecological animal geography. An authorized edition, rewritten and rev.; based on Tiergeographie auf Oekologischer Grundlage, by Richard Hesse. Ed. 2. New York, Wiley, 1951. 715 p.

C54 JEANNEL, R. G. La genèse des faunes terrestres; éléments de biogéographie. Paris, Presses Universitaires de France, 1942. 513 p.

C55 NEWBIGIN, M. I. Plant and animal geography. Ed. 3. New York, Dutton, 1950. 298 p.

C56 SCHMIDT, K. P. Faunal realms, regions, and provinces. In Q. Rev. Biol. 29(1954):322-331.
A review article, with bibliography, summarizing the literature of animal geography.

Periodical Lists

The list of journals abstracted in Biological abstracts (A11), usually published in the May issue each year, contains the more important zoological journals published throughout the world.

C57 ALLEN, F. P. A check list of periodical literature and publications of learned societies of interest to zoologists in the University of Michigan libraries. (U. of Mich. Mus. of Zool. C. no. 2.) Ann Arbor, Mich., University of Michigan Press, 1935. 83 p.

C58 APSTEIN, C. and WASIKOWSKI, K. Periodica zoologica; abkürzungsverzeichnis der wichtigsten zeitschriften-titel aus dem gebiet der zoologie und ihrer frenzgebiete. Leipzig, Akademische Verlagsgesellschaft, 1938. 82 p.

C59 MATTHES, E. Welche zeitschriften sind für ein zoologisches institut am wichtigsten. In Coimbra. U. Mus. Zool. Mem. e Estud. 122(1941):1-57

C60 UNDERWOOD, M. H. Bibliography of North American minor natural history serials in the University of Michigan libraries. Ann Arbor, Mich., University of Michigan Press, 1954. 197 p.

C61 ZOOLOGICAL SOCIETY of LONDON. List of the periodicals and serials in the Library . . . London, 1949. 186 p.

Economic Zoology 73

Textbooks

There are few textbooks on the general subject of economic zoology. Both of the following contain extensive bibliographies:

C62   REESE, A. M. Outlines of economic zoology. Ed. 4. Philadelphia, Blakiston, 1942. 359 p.
C63   WARDLE, R. A. Principles of applied zoology. London, Longmans, Green, 1929. 427 p.

ANIMAL HUSBANDRY

Animal husbandry is largely concerned with the breeding, selection, and production of livestock. The term "livestock" refers particularly to cattle, sheep, swine, goats, horses, and mules, though it may be extended to include certain other domesticated animals that have, or acquire, economic value. The pathology of livestock is considered under the heading "Veterinary medicine": dairy products and livestock after processing for food under "Food and nutrition": poultry and domesticated birds in "Poultry husbandry."

Abstracting Journals

Research literature of interest to the animal scientist will be found abstracted in a wide variety of journals. Biological abstracts (A11) covers extensively the literature on genetics and physiology, and some basic research on livestock breeding and feeding. Chemical abstracts (D8) can be used to survey publications on the chemistry of livestock products and chemical aspects of animal nutrition and physiology. The Experiment station record (A22) will continue to be useful because of its detailed abstracts and its coverage of years prior to the establishment of other abstracting services. The Bibliography of the literature of the minor elements and their relation to plant and animal nutrition (D27), largely taken from Chemical abstracts, brings together material pertaining to animal nutrition. The Nutrition abstracts and reviews (E13) covers world literature on the nutrition of man and animals, chemistry of foods, physiology of digestion, metabolism, growth, reproduction and resistance to disease, and the feeding of animals. Food science abstracts (E16) summarizes a great deal of literature on the use of livestock products, their processing, preservation, chemical composition, and analysis. Dairy science abstracts (E12) includes many abstracts of interest to animal scientists, particularly under the general heading "Husbandry." Resumptio genetica (B130) indexes literature on the genetics and breeding of livestock.

Many of the abstracting and indexing journals mentioned in the veterinary section will have references pertinent to animal husbandry, particularly on such topics as physiology, anatomy, biochemistry, nutritional disorders, reproduction, and metabolism.

A number of foreign journals have extensive indexing and abstracting sections. In the field of animal chemistry, genetics, and nutrition there was the Jahresbericht über die Fortschritte der Tierchemie (v. 1-49, 1871-1919) and its successor, Jahresbericht Physiologie und Experimentelle Pharmakologie (v. 1, 1920-date). An important bibliographical tool, particularly for the earlier years when it was virtually alone in the field, was the Jahresbericht für Agrikultur-Chemie (D10), which covered, in its annual lists of new books and journal articles, literature on animal production, feeds, nutrition, and animal husbandry. The Jahrbuch für Wissenschaftliche und Praktische Tierzucht (1906-date) contains review articles and extensive bibliographies arranged by broad subjects.

The list of journals with abstracting and indexing sections could be greatly extended (see Theodore Besterman's Index bibliographicus (A4). Most such journals, however, are intended as summaries of current literature, rather than as guides for retrospective literature searches.

C64   COMMONWEALTH BUREAU of ANIMAL BREEDING and GENETICS. Animal breeding abstracts. v. 1-   April 1933-   Edinburgh, 1933-

Issued quarterly. Author index in each issue. Annual author, geographical, and subject indexes. Abstracts are grouped under such subjects as breeding and genetics, reproduction, sex, growth and productivity of farm livestock, fur bearers, rodents, other mammals, poultry and other birds, etc. Book reviews and abstracts of reports of meetings, annual reports, etc., separately grouped at the back of each issue. About 2,000 abstracts were published in 1953. Particularly complete for British Commonwealth publications. Preceded by References to literature contained in periodicals received by the Imperial Bureau of Animal Genetics (nos. 1-20, 1931-1932).

C65   LANDWIRTSCHAFTLICHES zentralblatt. Abt. III: Tierzucht-Tierernährung. v. 1-   November 1955-   Berlin, Deutsche Akademie der Landwirtschaft, 1955-

Issued monthly or oftener. Abstracts, under broad headings, world literature on animal industry. Author index in each issue.

Bibliographies and Indexes

Most reports of research include lists of references that are frequently the most professional and best-defined guides to the literature on specific topics. The most complete index to such bibliographies as well as to books and journal articles that are primarily bibliographic, is the Bibliographic index (A2). The annual indexes of the Bibliography of agriculture (A24) and of the discontinued Experiment station record (A22) provide a key to lists of citations under the general heading "Bibliographies" and under specific topics. The Agricultural index (A23), because of its relative ease of use and its direct indexing under specific subjects, is a commonly used key to selected literature published in the United States. The following list is representative and selective rather than exhaustive.

C66   BIEN, C. Tropical beef cattle industry in the Western Hemisphere. (U.S.D.A. Bibliog. B. 19) 1952. 157 p.

C67   BROSCH, A. Die literatur über das schwein von 1538 bis zur gegenwart. Berlin, Pormetter, 1913. 96 p.

The Brosch bibliographies confine themselves chiefly to German literature.

C68   BROSCH, A. Das schrifttum über das rind. (Deut. Gesell. f. Züchtungskunde. Arb. Hft. 48.) Hannover, Schaper, 1930. 189 p.

Animal Husbandry

C69 DUERST, J. U. Grundlagen der rinderzucht. Berlin, Springer, 1931. 759 p.
    A bibliography of over 3,000 items that includes citations to literature published before 1850.

C70 GOOCH, D. W. Bibliography on the marketing of livestock, meat, and meat products. (U.S.D.A. Bibliog. B. 15) 1951. 209 p.
    English-language literature published from 1932-1950.

C71 HAWKS, E. B. Cattle, sheep, and goat production in the range country; a selected list of publications . . . (U.S.D.A. Libr. Bibliog. Contrib. 19) 1928. 78 p.

C72 IMPERIAL BUREAU of ANIMAL GENETICS. Bibliography on the biology of the fleece. Edinburgh, Oliver, 1931. 32 p.

C73 JACOBS, K. Livestock financing in the United States; selected references to material published 1915-1935. (U.S. Bur. Agr. Econ. Libr. Agr. Econ. Bibliog. 62) 1935. 57 p.
    Supersedes Agr. Econ. Bibliog. no. 7, 1925.

C74 NATIONAL RESEARCH COUNCIL, Canada. Bibliography on nutrition in sheep. (N. R. C. 1059) Ottawa, 1942. 86 p.

C75 NATIONAL RESEARCH COUNCIL, Canada. Bibliography on the influence of mineral deficiencies on growth and reproduction of farm animals; prep. by M. E. Whalley. Ottawa, 1937. 89 p.

C76 NATIONAL RESEARCH COUNCIL, Canada. Bibliography on wool. Ottawa, 1927. 174 p.

C77 OSTERTAG, R. Bibliographie der fleischbeschau. Stuttgart, Enke, 1905. 446 p.

C78 PLUMB, C. S. A partial index to animal husbandry literature. Columbus, Ohio, 1911. 94 p.
    Includes annotated references to early literature.

C79 PLUMB, C. S. Registry books on farm animals: a comparative study. Columbus, Ohio, Ohio State University Press, 1930. 306 p.
    A bibliography of herd, stud, and flock books, with historical and descriptive information.

C80 RENNER, F. G. A selected bibliography on management of western ranges, livestock, and wildlife. (U.S.D.A. Misc. P. 281) 1938. 468 p.

C81 SMITH, A. D. B. Genetics of cattle. In Bibliog. Genet. 10(1933):1-104.

C82 THOMPSON, G. F. Index to literature relating to animal industry in the publications of the U.S. Department of Agriculture, 1837-1898. (U.S.D.A. Div. of Pub. B. 5) 1900. 676 p.

C83 U.S. SUPERINTENDENT of DOCUMENTS. Animal industry, farm animals, poultry, and dairying. (Price List 38) 1954. 11 p.
    Revised approximately annually.

Encyclopedic Works

V. 3 of Bailey's Cyclopedia of American agriculture (A47) contains summary information on breeds and breed societies, with bibliographic references to older literature.

C84 JOHANNSON, I. Husdjursraserna: hästar, nötkreatur, får och getter, svin, kaniner och pälsdjur, fjäderfä. Stockholm, LTs Förlag, 1953. 523 p.

C85 MASON, I. L. World dictionary of breeds, types, and varieties of livestock. (Commonwealth Bur. of Anim. Breeding and Genet. Tech. Commun. no. 8.) Slough, Commonwealth Agricultural Bureaux, 1951. 272 p.
Pt. I lists names of livestock, including synonymous names, place of origin, present distribution, economic use, relationship to other breeds, description of breed characteristics, and the origin of the name. Pt. II is arranged by country, showing principal breeds and relative numbers, with bibliographic references to literature of the country and the breeds.

C86 SANDERS, A. H. The cattle of the world; their place in the human scheme - wild types and modern breeds in many lands. Washington, National Geographic Society, 1926. 142 p.
A popular presentation with colored illustrations.

C87 SCHMID, A. Rassenkunde des rindes. Bern, Benteli, 1942. 2 v.
V. 1 describes breeds of cattle by country; v. 2 contains illustrations of breeds of the world.

C88 U.S. BUREAU of FOREIGN COMMERCE. Cattle and dairy farming. (U.S. Consular Rpt., House Exec. Doc. no. 51, 49th Congress, 1st Session.) Washington, 1888. 2 v.
Encyclopedic information on dairy cattle of the world.

Dictionaries

A glossary of terms used in the field of animal husbandry, presumably polyglot, has been proposed by a joint group of the F.A.O. Agricultural Division and the European Association of Animal Production. No publication plans are known as of this writing. Most of the dictionaries listed in A53-A61 include terms from the field of animal husbandry. Animal geneticists, particularly, will find Knight (B149) useful. Davis' Dictionary of dairying (E77) contains many terms common to the fields of dairying and animal husbandry.

Other dictionaries in this and allied fields are listed in the Bibliography of interlingual scientific and technical dictionaries (A53). Dictionaries will be found listed in the annual subject indexes of the Bibliography of agriculture (A24) under appropriate headings and subheadings.

C89 BARANOWSKI, Z. The international horseman's dictionary. English-French-German. London, Museum Press, 1955. 192 p.

C90 BRIQUET, R. Pequene dicionário inglês-português de têrmos empregados em anatomia, ezoognósia, fisiologia, zootecnia e tecnologia dos produtos animais. Ed. 2. (Brazil. Serv. de Inform. Agr. Sér. Didática no. 10.) Rio de Janeiro, 1952. 166 p.

C91 INTER-AMERICAN INSTITUTE of AGRICULTURAL SCIENCES. Diccionario Ingles-Espanol de termines de ganaderia. Turrialba, 195?. 13 p.

C92 BAGUÉ, J. Glosario de biologia animal (Español-Inglés). In Puerto Rico. Departamento de Agricultura y Comercio. Alamanaque agricola de Puerto Rico, 1951-1952. p. 195-373.

C93 BAGUÉ, J. Glossary of animal biology (English-Spanish). In Puerto Rico. Departamento de Agricultura y Comercio. Almanaque agricola de Puerto Rico, 1950. p. 143-309.

C94   SHIBATA, S. Saishin hikkei chikusan soten. Tokyo, Asakura, 1954.
      506 p.
      A dictionary of livestock.

Directories

Because of the many subject fields associated with animal husbandry, the potential sources of information concerning organizations and persons are extremely varied. Lists of organization members appear in journals, in proceedings, in biographical series, in trade journals, and in directories of government workers. Some of these are published in a consistent series, others in irregular issues. Keeping a reference collection of this type of information current and usable requires constant attention to publication notices in journals and to current listings in bibliographies. Particularly useful are listings under the heading "Organizations - Directories," and the subheading "Directories" in the annual indexes of the Bibliography of agriculture (A24) and under the many subheadings of the main heading "Directories" in the Bulletin of the Public Affairs Information Service.
Winchell (A8) describes many of the guides to international organizations. Agricultural organizations with international membership are listed and described in UNESCO's Directory of international scientific organizations (A126).
The County agents' directory (A66) includes lists for the United States of national registry associations, livestock associations, fairs, and expositions. Davis (A67) provides a convenient list of directories for the meat and dairy industries. The journal Better farming methods, in its November issue each year, publishes directories of breed associations, farm organizations, and agricultural personnel of various agencies. Ensminger (C100) has an extensive list of breed registry associations, as have many textbooks in the field. The Breeder's gazette publishes frequent livestock association directories (e.g., March, 1955, issue). Animal research organizations are listed in the National Research Council's Handbook of scientific and technical societies (A125) and national animal husbandry organizations of various kinds in Judkins (A123).
Federal government agencies issue lists of livestock registry associations. The U.S. Bureau of Dairy Industry, in its BDI-Inf series, publishes a number of directories such as, Associations of breeders of registered dairy cattle (BDI-Inf-6), and Address list of national and international dairy associations (BDI-Inf-4). The U.S. Animal Industry Bureau publishes American beef and dual-purpose cattle breeders' associations (AHD no. 4) and American breed associations: swine (AHD no. 6).
Similar directories are published in many foreign countries. For England, there is the Farmer's yearbook encyclopaedia and directory of pedigree breeders; Australia has the Australian pastoral directory; list of stockowners. There are regional directories, such as the Northwest livestock directory and state directories, such as the Oklahoma livestock breeders directory, issued by the Oklahoma Department of Agriculture.
In addition to directories of associations and individuals, animal husbandmen work with a third type of directory: lists of livestock. These are published as herdbooks (cattle and swine), studbooks (horses), flock books (sheep), special lists of sires, and advanced registry animals. The most comprehensive current listing of these publications appears in the

Bibliography of agriculture's (A24) annual indexes under such headings as "Cattle - Herdbooks," "Swine - Herdbooks," and "Horses - Studbooks." Most of the domestic registry associations are members of the National Society of Livestock Record Associations, and information concerning individual breed associations and their publications can be obtained either from the association itself or from the National Society. These animal registry books are treated historically and listed in Plumb's Registry books on farm animals (C79), in Bailey (A47), and in many of the textbooks on livestock, such as Vaughan (C105), Briggs (C99), and Ensminger (C100).

In addition to the standard biographical directories, particularly American men of science (A83), individuals in the field of animal husbandry are listed in membership directories of associations, in trade directories, and in government lists.

C95   FOOD and AGRICULTURE ORGANIZATION of the UNITED NATIONS. Index of agricultural research institutions in Europe concerned with animal production and their principal lines of investigation. Rome, 1953. 151 p.

C96   INTERNATIONAL directory of pedigree stock breeders. London, Vernon, 1926-1930/31. 3 v.

C97   INTERNATIONAL INSTITUTE of AGRICULTURE. Les institutions de zootechnie dans le monde. International directory of animal husbandry institutions. Rome, 1933. 325 p.

C98   U.S. AGRICULTURAL RESEARCH SERVICE. Working reference of livestock regulatory establishments, stations, and officials. October 1954-   Washington, 1954-
This supersedes earlier directories of the Bureau of Animal Industry, which had been issued continuously since 1906. It includes establishments operating under Federal Meat Inspection, public stockyards, licensed manufacturers of biological products, pathological laboratories, state officials in charge of animal disease control, Washington and field personnel, and veterinarians.
For other government directories, see the annual indexes of the Bibliography of agriculture (A24) under such headings as "Livestock - Directories," "Meat - Directories," "Swine - Directories," etc. Also consult the general directories of government personnel listed in A75-A79.

Handbooks, Texts, etc.

For a selected list of textbooks and journals in the field, see the U.S. Department of Agriculture Library's Selected list of American agricultural books and periodicals (A39).

General information on breeds is contained in a number of Farmers' bulletins, which are revised frequently, such as Beef cattle breeds for beef and for beef and milk (no. 1779), Dairy cattle breeds (no. 1443), Breeds of draft horses (no. 619), etc.

C99   BRIGGS, H. M. Modern breeds of livestock. New York, Macmillan, 1949. 772 p.

C100   ENSMINGER, M. E. The stockman's handbook. Danville, Ill., Interstate, 1955. 598 p.

General information on all aspects of livestock management, including laws, lists of associations, magazines, books, weights and measures, etc.

C101 HOGNER, D. C. Farm animals and working and sporting breeds of the United States and Canada. London, Oxford, 1945. 194 p.

C102 MORRISON, F. B. Feeds and feeding, a handbook for the student and stockman. Ed. 21. Ithaca, N. Y., 1948. 1,207 p.
Contains encyclopedic information on feeding, with extensive bibliographies.

C103 NATIONAL RESEARCH COUNCIL. COMMITTEE on ANIMAL NUTRITION. Recommended nutrient requirements for domestic animals. no. 1-   June 1944-   Washington, 1944-
Revised irregularly as needed. Nine numbers issued to date: no. 1, Poultry; no. 2, Swine; no. 3, Dairy cattle; no. 4, Beef cattle; no. 5, Sheep; no. 6, Horses; no. 7, Foxes and Minks; no. 8, Dogs; no. 9, Rabbits. Each number has a bibliography.

C104 PLUMB, C. S. Types and breeds of farm animals. Boston, Ginn, 1920. 820 p.

C105 VAUGHAN, H. W. Breeds of live stock in America. Columbus, Ohio, Adams, 1931. 780 p.

History - General

Historical writing in the field of animal husbandry is abundant and varied. The Bibliography of agriculture (A24) lists much of it under such headings as "Cattle - History," "Livestock - History," "Horses - History," etc., in the annual indexes. General sketches with bibliographical citations can be found in Briggs (C99), Vaughan (C105), Duerst (C69, C110), etc. The histories of most breeds of livestock have been published by breed societies. Frequently, the first volume of a herd, flock, or studbook will contain historical material on a breed.

C106 ALLEN, L. F. American cattle, their history, breeding, and management. New York, Taintor, 1868. 528 p.

C107 BARANSKI, A. Die vorgeschichtliche zeit im lichte der hausthiercultur. Wien, Perles, 1879. 296 p.
A history of livestock in antiquity.

C108 CLAWSON, M. The western range livestock industry. New York, McGraw-Hill, 1950. 401 p.

C109 CLEMEN, R. A. American livestock and meat industry. New York, Ronald, 1923. 872 p.

C110 DUERST, R. J. Die rinder von Babylonien, Assyrien, und Ägypten und ihr zusammenhang mit den rindern der alten welt. Berlin, Reimer, 1899. 94 p.
Additional references can be found in Duerst's Grundlagen der rinderzucht (C69).

C111 FEIGE, E. Die haustierzonen der alten welt. (Ergänzh. no. 198 zu Petermanns Geographische Mitteilungen.) Gotha, Perthes, 1928. 121 p.
Extensive bibliography, p. 117-121.

C112 HOUCK, U. G. The Bureau of Animal Industry of the United States Department of Agriculture; its establishment, achievements,

and current activities. Washington, 1924. 390 p.
C113 KAHN, H. Records in the National Archives relating to the range cattle industry, 1865-1895. In Agr. Hist. 20(1946):187-90.
C114 LAWRENCE, J. M. American livestock biographies. Roseville, Calif., 1951. 393 p.
History and description of the common breeds of livestock in America, with some information about registry associations, etc.
C115 LEAVITT, C. T. Attempts to improve cattle breeds in the United States, 1790-1860 ... Chicago, 1933. p. 51-67. (Reprinted from Agr. Hist. 7(1933):51-67.)
C116 LOW, D. The breeds of domestic animals of the British Islands. London, Longman, 1842. 2 v.
V. 1, The horse and ox. V. 2, Sheep, goats, and hogs.
C117 MALIN, D. F. The evolution of breeds; an analytical study of breed building as illustrated in Shorthorn, Hereford, and Aberdeen Angus cattle, Poland China and Duroc Jersey swine. Des Moines, Wallace, 1923. 278 p.
C118 MOLL, L. La connaissance générale du boeuf, études de zootechnie pratique sur les races bovines de la France, de l'Algérie, de l' Angleterre, de l' Allemagne, de la Suisse, de l' Autriche, de la Russie et de la Belgique, avec une atlas de 83 figures. Paris, Didot, 1860. 600 p.
C119 MORSE, E. W. The ancestry of domesticated cattle. In U.S. Bur. of Anim. Ind. Ann. Rpt., 27th, 1910, p. 187-239. 1912.
The bibliography, p. 233-239, is noteworthy.
C120 POWELL, F. W. The Bureau of Animal Industry, its history, activities, and organization. (Inst. for Govt. Res. Serv. Monog. of the U.S. Govt. no. 41.) Baltimore, Johns Hopkins Press, 1927. 190 p.
C121 PRENTICE, E. P., and others. American dairy cattle; their past and future; with chapters on dairy cattle in America. New York, Harper, 1942. 453 p.
Historical material on European backgrounds and American history of breeds, with bibliographical citations.
C122 SALMON, D. E. The United States Bureau of Animal Industry, at the close of the nineteenth century, 1884-1900. Washington, 1901. 387 p.
C123 STORER, J. The wild white cattle of Great Britain. An account of their origin, history, and present state. London, Cassell, 1879. 384 p.
C124 THOMPSON, J. W. A history of livestock raising in the United States, 1607-1860. (U.S. Bur. Agr. Econ. Agr. Hist. Ser. 5) 1942. 182 p.
Traces livestock raising back to its European beginnings. Has an extensive bibliography.
C125 WHITEHEAD, G. K. The ancient white cattle of Britain and their descendants. London, Faber, 1953. 174 p.
C126 WILSON, J. The evolution of British cattle and the fashioning of breeds. London, Vinton, 1909. 147 p.
C127 YOUATT, W. Cattle: their breeds, management, and diseases. London, Baldwin, 1834. 600 p.
Contains early history of breeds.

Animal Husbandry

History - Cattle - Individual Breeds

Aberdeen-Angus
C128 AMERICAN ABERDEEN-ANGUS BREEDERS' ASSOCIATION. Origin of the Aberdeen-Angus and its development in Great Britain and America. Chicago, 1914. 36 p.
Bibliography, p. 34-36.
C129 MACDONALD, J. and SINCLAIR, J. History of Aberdeen-Angus cattle. London, Vinton, 1910. 682 p.
C130 SANDERS, A. H. A history of Aberdeen-Angus cattle, with particular reference to their introduction, distribution, and rise to popularity in the field of fine beef production in North America. Chicago, New Breeder's Gazette, 1928. 1,042 p.
C131 SINCLAIR, J. History of polled Aberdeen or Angus cattle. London, Blackwood, 1882. 459 p.

Guernsey
C132 HILL, C. L. The Guernsey breed. Waterloo, Ia., Kimball, 1917. 417 p.
C133 PRENTICE, E. P. The history of Channel Island cattle, Guernseys and Jerseys. Williamstown, Mass., Mount Hope Farm, 1940. 102 p.

Hereford
C134 HAZELTON, J. M. History and hand book of Hereford cattle and Hereford bull index. Kansas City, Hereford Journal, 1925. 391 p.
C135 MACDONALD, J. and SINCLAIR, J. History of Hereford cattle. London, Vinton, 1909. 501 p.
C136 MILLER, T. L. History of Hereford cattle ... with which is incorporated a history of the Herefords in America, by Wm. H. Sotham. Chilliocothe, Mo., Sotham, 1902. 592 p.
C137 SANDERS, A. H. The story of the Herefords. Chicago, Breeder's Gazette, 1914. 1,087 p.
C138 WILLHAM, O. S. A genetic history of the Hereford breed of cattle in the United States. In J. of Hered. 28(1937):283-294.

Holstein-Friesian
C139 PRESCOTT, M. S. Holstein-Friesian history. Lacona, N. Y., Corse Press, 1930. 254 p.

Jersey
C140 BOSTON, E. J. Jersey cattle. New York, Van Nostrand, 1955. 232 p.
C141 GOW, R. M. The Jersey; an outline of her history during two centuries - 1734-1935. New York, American Jersey Cattle Club, 1936. 539 p.

Shorthorn
C142 ALLEN, L. F. History of the Short-horn cattle: their origin, progress, and present condition. Ed. 2. Buffalo, N. Y., 1883. 280 p.
C143 SANDERS, A. H. Red, white, and roan; stories relating to the origin of the Shorthorn breed of cattle in Great Britain; its early introduction into America, and its growth in popularity throughout the United States in recent years. Chicago, American Shorthorn Breeders' Assoc., 1936. 630 p.

C144 SANDERS, A. H. Shorthorn cattle. Chicago, 1918. 1,021 p.
C145 SINCLAIR, J. History of shorthorn cattle. London, Vinton, 1907. 895 p.

History - Sheep

C146 CONNOR, L. G. A brief history of the sheep industry in the United States. In Amer. Hist. Assoc. Ann. Rpt. ... for the year 1918. Washington, 1921, p. 89-197.
C147 SALMON, D. E. Special report on the history and present condition of the sheep industry of the United States. Washington, U.S. Bureau Animal Industry, 1892. 1,000 p.
C148 TOWNE, C. W. and WENTWORTH, E. N. Shepherd's empire. Norman, University of Oklahoma Press, 1945. 364 p.
   History of sheep industry in the United States. Extensive bibliography.
C149 WENTWORTH, E. N. America's sheep trails: history, personalities. Ames, Ia., Iowa State College Press, 1948. 667 p.
C150 YOUATT, W. Sheep: their breeds, management, and diseases. London, Baldwin and Cradock, 1837. 568 p.

History - Swine

C151 DAVIS, J. R. and DUNCAN, H. S. History of the Poland China breed of swine. Omaha, Poland China Historical Association, 1921. 2 v.
C152 TOWNE, C. W. and WENTWORTH, E. N. Pigs, from cave to corn belt. Norman, University of Oklahoma Press, 1950. 305 p.
   Bibliography, p. 279-290.

History - Horses

C153 RIDGEWAY, W. The origin and influence of the thoroughbred horse. Cambridge, Eng., Cambridge University Press, 1905. 538 p.
C154 SANDERS, A. H. A history of the Percheron horse. Chicago, Breeder's Gazette, 1917. 602 p.
C155 VERNON, A. The history and romance of the horse. New York, Dover, 1946. 525 p.
C156 YOUATT, W. A history of the horse. Washington, Green, 1834. 360 p.

Geography

Maps, giving such livestock information as the distribution and population of farm animals, can be found in many of the general economic atlases and in agricultural atlases such as Van Royen (A107), Bacon (A100), and Finch (A106). The Atlas of American agriculture (A105) in its final form did not include livestock, but much of the data on farm animals assembled for it was printed as a section of the 1921 Yearbook of agriculture (A81), entitled "Graphic summary of American agriculture." This "graphic summary" was continued in references C157, C158, and C162.

Many Census Bureau publications contain graphic livestock information. Hainsworth (A102) summarizes world livestock figures graphically.

C157 BAKER, O. E. A graphic summary of American agriculture based largely on the census. (U.S.D.A. Misc. P. 105) 1931. 228 p.
C158 BAKER, O. E. A graphic summary of farm animals and animal products (based largely on the census of 1930 and 1935). (U.S. D.A. Misc. P. 269) 1939. 88 p.
C159 INTERNATIONAL INSTITUTE of AGRICULTURE. Atlas international zootechnique. Rome, Bestetti & Tumminelli, 1930. v. 1, pts. 1-4. Not completed. Intended to provide descriptions of cattle breeds of each country, with maps of distribution. Completed maps for Germany, Hungary, Netherlands, and Switzerland.
C160 NEWBIGIN, M. I. Plant and animal geography. Ed. 3. New York, Dutton, 1950. 298 p.
C161 SAUER, C. O. Agricultural origins and dispersals. (Bowman Mem. Lectures, ser. 2.) New York, American Geographical Society, 1952. 110 p.
C162 U.S. BUREAU of AGRICULTURAL ECONOMICS. A graphic summary of farm animals and animal products (based largely on the census of 1940). (U.S.D.A. Misc. P. 530) 1943. 88 p.
C163 VEYRET, P. Géographie de l'élevage. (Géog. Humaine, 23.) Paris, Gallimand, 1951. 254 p.

Periodical Lists

The most extensive current list of journals in the field is published at frequent intervals in Animal breeding abstracts (C64) (for instance, see the list published with the annual index to v. 22, 1954). Apstein's Periodica zoologica (C58) is an older list and includes animal husbandry journals. There is an extensive list of breed journals in Ensminger (C100). See also lists described in A109-A122.

Miscellaneous

C164 HAMMOND, J., ed. Progress in the physiology of farm animals. v. 1- London, Butterworth, 1954-
Issued annually. Summary articles with bibliographies.
C165 PROSE and poetry of the live stock industry of the United States. Denver, National Live Stock Historical Association, 1905. 757 p.

POULTRY HUSBANDRY

The term "poultry" as used here includes chickens, turkeys, guineas, ducks, geese, pigeons, pheasants, and other domesticated birds that have marketable value. Poultry husbandry includes the selection, breeding, production, housing, management, marketing, processing, and distribution of poultry and poultry products.

84   C. Animal Sciences

Abstracting Journals

The field lacks a comprehensive indexing or abstracting journal. The scope of the literature is indicated by the fact that, during 1953, the Bibliography of agriculture (A24) listed over 3,100 items under the general heading "Poultry." Biological abstracts (A11) cited and abstracted over 500 references during that year. Three British abstracting journals, which cover literature dealing with one phase or another of this subject, cited or abstracted about 600 publications in the same period: Animal breeding abstracts (C64) about 250 on the production of poultry, production of eggs, growth, genetics, physiology of the egg, reproduction and sex; Nutrition abstracts and reviews (E13) about 175 publications on the feeding and nutrition of poultry and the chemical composition of poultry products; Food science abstracts (E16) about 170 abstracts on the processing and composition of poultry products as food. There is some duplication among the British journals, but the abstracts appear to be written independently for each. The most recent literature is cited in the Bibliography of agriculture: Biological abstracts requires six months to several years to abstract any one publication, depending on language and accessibility; the British journals are somewhat slower. Literature on the chemical composition and analysis of poultry products and poultry feeding stuffs is abstracted in Chemical abstracts (D8). During the years it was issued, 1889-1946, the Experiment station record (A22) published abstracts of world literature on various phases of poultry husbandry and research.

The abstracting service offered by the World's poultry science journal (1945-date, Ithaca, N. Y.) and its predecessor, the International review of poultry science (v. 1-13, no. 1/2, 1928-1940, Rotterdam) is not taken into consideration here because of inadequate indexing and because most of these abstracts are reprinted from other abstracting services. The Institute of American Poultry Industries for years maintained an abstract file on cards. For the period 1911-1938, this file included about 5,500 abstracts. From September 1938 to December 1944, these abstracts were published in the U.S. Egg and poultry magazine.

Bibliographies - General

Systematic literature and literature dealing with fowl as birds, rather than as economic products, is indexed in the Zoological record (C10). For literature in this category, other bibliographies cited in the section on economic zoology should also be considered.

C166 STRONG, R. M. A bibliography of birds, with special reference to anatomy, behavior, biochemistry, embryology, pathology, physiology, genetics, ecology, aviculture, economic ornithology, poultry culture, evolution, and related subjects ... (Field Mus. of Nat. Hist. Zool. Ser. v. 25, pts. 1-3) Chicago, Field Museum Press, 1939-1946. 3 v.
A basic work in the field of zoology that links the literature of economic fowl to the broader field of avian literature. Pts. 1-2, Authors; pt. 3, Subject index. Chiefly cites literature published before 1926, to coincide with the first volume of Biological abstracts (A11). Poultry economics and management included as

subjects, but not extensively represented by literature. Long list of ornithological and poultry periodicals and abstract and index guides to literature on birds in pt. 1. Pt. 3 includes an extensive list of ornithological bibliographies and a list of periodicals containing abstracts and reviews of avian literature.

Bibliographies - Early

Poultry literature is historically considered in the following four publications.

C167 JEFFERS, F. H. A selection of old poultry books, English and American, with a chronology of significant dates in poultry history; and A digest of R. C. Punnett's "Notes on Old Poultry Books." Windsor, N. Y., Windsor Standard, 1954. 26 p.

C168 PAYNE, L. F. List of poultry books. (Kans. Agr. Exp. Sta. Dept. of Poultry Husb. Contrib. no. 147) Manhattan, Kans., Kansas State College Press, 1943. 83 p.

C169 PAYNE, L. F. Old poultry books. In World's Poultry Sci. J. 2(1946): 116-121.

C170 PUNNETT, R. C. Notes on old poultry books. London, Feathered World, 1930. 40 p.

Bibliographies - Special

The best sources of literature lists on specific subjects are research articles and monographs. Those with particularly noteworthy bibliographies are indexed under appropriate subject headings in such guides as the Bibliographic index (A2), in the subject indexes to the Bibliography of agriculture (A24), and the Experiment station record (A22). Of these specialized bibliographies, mention should be made of the very extensive literature lists in Romanoff's The avian egg (C209), of the more than 3,000 references in Ewing's Poultry nutrition (C206), and of the highly specialized lists appended to practically all articles in Poultry science.

The Poultry Division, Agricultural Marketing Service, U.S. Department of Agriculture, issues and revises periodically a large number of short bibliographies on specialized subjects such as farm egg-handling practices, egg candling and grading, interior quality of eggs, egg-shell texture, egg size, washing eggs, oil-protecting eggs, egg storage, merchandising eggs, nutritive value of eggs, frozen and dried eggs, poultry processing, freezing and storage, and poultry merchandising.

The following list of bibliographies on various phases of poultry husbandry is selected and typical, rather than exhaustive.

C171 BERCAW, L. O. The poultry industry, a selected list of references on the economic aspects of the industry. 1920-1927. (U.S. Bur. of Agr. Econ. Agr. Econ. Bibliog. 24) 1928. 106 p.
Chiefly confined to literature of the United States and Canada. Periodical articles omitted.

C172 FABER, F. L. Bibliography on poultry and egg marketing; a selected list of references on the economic aspects of poultry and egg marketing in the United States and Canada. Boston, New England Research Council on Marketing and Food Supply, 1953. 162 p.

C173 HAWKS, E. B. A bibliography of poultry. (U.S.D.A. Libr. B. 18) 1897. 32 p.

Includes books, bulletins, and a list of poultry journals. Subject index.

C174 HENNEFRUND, H. E. Bibliography on the poultry industry in countries other than Canada and the United States. (U.S.D.A. Libr. Libr. List 12) 1944. 113 p.

C175 PAYNE, L. F. Poultry theses, presented for advanced degrees at land grant colleges and universities in the United States of America, 1896-1950. (Contrib. no. 197 from the Dept. of Poultry Husb., Kansas State College, Manhattan. P. no. 1 of the James E. Rice Poultry Libr., Cornell University.) Ithaca, N. Y., James E. Rice Memorial Poultry Library, Cornell University, 1952. 106 p.

Lists masters' theses, as well as doctoral dissertations. Doctoral dissertations on poultry are listed in the annual Doctoral dissertations accepted by American universities (New York, Wilson, 1934-   ).

C176 U.S. BUREAU of AGRICULTURAL CHEMISTRY and ENGINEERING. Poultry housing, a selected list of references. Washington, 1941. 49 p.

C177 U.S. BUREAU of AGRICULTURAL ECONOMICS. LIBRARY. Egg auctions. (Econ. Libr. List 4) 1939. 20 p.

C178 U.S. DEPARTMENT of AGRICULTURE. Bibliography of poultry publications, comp. by the Departmental Poultry Committee. Washington, 1938. 76 p.

C179 U.S. OFFICE of EXPERIMENT STATIONS. Bibliography on poultry husbandry. Chickens, breeding, egg investigations, feeds and nutrition, meat investigations, and physiology, 1949 and 1950, comp. by R. B. Nestler. Washington, 1951. 54 p.

C180 U.S. OFFICE of EXPERIMENT STATIONS. Bibliography on poultry industry; chickens, turkeys, aquatic fowl, gamebirds, comp. by R. B. Nestler. v. 1-   1948/50-   Washington, 1951-

C181 U.S. SUPERINTENDENT of DOCUMENTS. Animal industry, farm animals, poultry, and dairying. (Price List 38) 1954. 11 p. Revised approximately annually.

Bibliographies - United States Experiment Station
and Extension Service Publications

Experiment station and extension publications on poultry are listed in the following special bibliographies. Current lists of these publications now appear in the Bibliography of agriculture (A24) and are indexed in the Agricultural index (A23). See also the general lists of government publications cited in (A36-A44).

C182 CARD, L. E. A check list of poultry bulletins. In Poultry Sci. 11(1932):339-353.

Lists 496 titles issued by agricultural experiment stations of the

United States from their establishment to the end of 1930. Includes regular bulletins, research and technical bulletins, but excludes popular bulletins and annual reports. Continued by (C183).
C183 CARD, L. E. A check list of poultry bulletins, 1931-1940. In Poultry Sci. 21(1942):58-69.
C184 NEW YORK. AGRICULTURAL EXPERIMENT STATION, Ithaca. Extension publications on poultry science issued by the agricultural experiment stations of the United States ending December, 1942. Rev. Ithaca, N. Y., 1943. 144 p.
C185 ROMANOFF, A. L. Popular publications in the field of poultry industry, issued by the agricultural experiment stations of the United States ending June, 1941. Ithaca, N. Y., Cornell University Agricultural Experiment Station, 1942. 111 p.

Encyclopedias

C186 ENCYCLOPAEDIA of poultry . . . Including the latest edition of the Poultry Club standards. London, Waverley, 1935. 2 v.
C187 SEIDEN, R. Poultry handbook, an encyclopedia for good management of all poultry breeds. Ed. 2. New York, Van Nostrand, 1952. 444 p.
Contains several thousand articles, ranging from definitions to lengthy discussions.

Dictionaries

Definitions of poultry terms are included in several of the encyclopedic-type handbooks. Such handbooks as Seiden (C187), the American standard of perfection (C190), and many of the textbooks, contain glossaries and can serve as dictionaries of the field. Many of the dictionaries listed in A53-A61 also define poultry science terms.

C188 LISSOT, G. Dictionnaire d'aviculture, suivi d'un lexique avicole Anglais-francais. Paris, Flammarion, 1948. 109 p.
C189 SUTTIE, D. F. Dictionary of poultry. London, Blackie, 1929. 280 p.

Breeds and Species

A number of Farmers' bulletins, published by the U.S. Department of Agriculture, give specialized information on various breeds: no. 2065, Breeds of chickens for meat and egg production; no. 767, Goose raising; no. 2066, Ornamental and game breeds of chickens; no. 1391, Guinea fowl; and others.
A great deal of information is distributed through the various agencies of the National Poultry Improvement Plan and the National Turkey Improvement Plan (see U.S. Department of Agriculture Misc. P. no. 300 and revisions, and Misc. P. no. 555 and revisions).

C190 AMERICAN POULTRY ASSOCIATION. The American standard of perfection. Atlanta, 1953. 585 p.
The standard descriptive guide to all recognized varieties of fowl.

First issued in 1874, new editions are published at irregular intervals. "Glossary of technical terms," p. 17-43.

C191 BRITISH poultry standards; complete specifications and judging points of all standardized breeds and varieties of poultry as compiled by the specialist breed societies and recognised by the Poultry Club of Great Britain. London, Poultry World, 1954(?). 348 p.
The British version of the American standard of perfection (C190). Supersedes Poultry club standards of perfection, 1930.

C192 JULL, M. A. The races of domestic fowl. In Natl. Geog. Mag. 51(1927):379-452.
A popular, illustrated account.

C193 PAYNE, L. F. and AVERY, T. B. International poultry guide for flock selection; covering the more popular breeds and varieties of chickens and turkeys. Ed. 2. Kansas City, Mo., International Baby Chick Association, 1950. 247 p.
Intended to supplement the American standard of perfection (C190).

C194 BEEBE, C. W. A monograph of the pheasants. London, Witherby, 1918-1922. 4 v.

C195 FRIEDMANN, H. The birds of North and Middle America. Part X: Family Cracidae . . . (Smithsonian Inst. U.S. Natl. Mus. B. 50, Part X) 1946. 484 p.
Includes grouse, partridges, pheasants, guinea fowl, turkeys, etc.

C196 LEVI, W. M. The pigeon. Columbia, S. C., Bryan, 1941. 512 p.
An encyclopedic work, including about 1,000 bibliographic references.

C197 PETERS, J. L. Check-list of birds of the world. Cambridge, Mass., Harvard University Press, 1931-1945. 5 v.

C198 PHILLIPS, J. C. A natural history of the ducks. Boston, Houghton Mifflin, 1922-1926. 4 v.
Extensive bibliography, v. 4, p. 321-447.

Directories

Membership lists are frequently available from associations such as the Poultry Science Association, American Poultry Association, and the Institute of American Poultry Industries. The Turkey world has special directory issues during the year, listing clubs, organizations, products, hatcheries, and breeders. Specialized directories of interest to poultry producers also appear regularly in such journals as Quick frozen foods (E120), Hatchery tribune and feed retailer, and Hatchery and feed.

Publications of this type are listed regularly under the heading "Poultry - Directories" in the Bibliography of agriculture's (A24) annual subject indexes. The following titles are typical of the many individual directories and directory series that are issued irregularly, under varying titles, and by different agencies. (See note on directories, p. 15.)

C199 U.S. AGRICULTURAL RESEARCH SERVICE. Directory of U.S.R.O.P. breeders qualifying for the U.S. Register of Merit Breeding Stage, 1953-1954. (ARS 53-1, rev.) 1955. 12 p.
Lists U.S. Register of Performance poultry breeders.

C200 U.S. AGRICULTURAL RESEARCH SERVICE. ANIMAL and POULTRY HUSBANDRY RESEARCH BRANCH. Hatcheries and dealers

participating in the National Poultry Improvement Plan. (ARS 53-6) 1955. 82 p.
C201 U. S. AGRICULTURAL RESEARCH SERVICE. ANIMAL and POULTRY HUSBANDRY RESEARCH BRANCH. Hatcheries, dealers, and independent flocks participating in the National Turkey Improve-Plan, 1956. (ARS 53-8, rev.) 1956. 33 p.
C202 U. S. BUREAU of ANIMAL INDUSTRY. Poultry specialty clubs and national associations. (A.H.D. 18) 1953. 4 p.
C203 U. S. PRODUCTION and MARKETING ADMINSTRATION. List of plants approved under the poultry inspection and grading program. Washington, 1952. 25 p.
C204 U. S. AGRICULTURAL MARKETING SERVICE. Directory, names and addresses of personnel who may be contacted for grading and/or inspection service on poultry and egg products and domestic rabbits. Washington, 1954. 9 p.
C205 WHO'S who in the egg and poultry industries, 1954. New York, Urner-Barry, 1954. 442 p.
Issued annually. Includes lists of cooperatives, hatcheries, processors, warehouses, shippers, suppliers and supplies, etc.

Textbooks

There are a number of specialized treatises that are encyclopedic in coverage of their subjects.

C206 EWING, W. R. Poultry nutrition. Ed. 4. South Pasadena, Calif., 1951. 1,518 p.
Over 3,000 bibliographic references at ends of chapters.
C207 HUTT, F. B. Genetics of the fowl. New York, McGraw-Hill, 1949. 590 p.
C208 PATTEN, B. M. Early embryology of the chick. Ed. 4. Philadelphia, Blakiston, 1951. 244 p.
C209 ROMANOFF, A. L. and A. J. The avian egg. New York, Wiley, 1949. 918 p.
Bibliography, p. 809-872.
C210 STURKIE, P. D. Avian physiology. Ithaca, N. Y., Comstock, 1954. 423 p.
C211 TAYLOR, L. W. Fertility and hatchability of chicken and turkey eggs. New York, Wiley, 1949. 423 p.

Statistics

C212 U. S. BUREAU of the CENSUS. Fifteenth census of the United States: 1930. Agriculture. Chickens and chicken eggs and turkeys, ducks, and geese raised on farms. Chickens and poultry products, with selected items by size of flock, for the United States, states, and counties, 1930 and 1929. Washington, 1933. 563 p.
C213 U. S. BUREAU of the CENSUS. Sixteenth census of the United States: 1940. Agriculture. Special poultry report. Statistics by geographic divisions and states for poultry of all kinds on hand and raised; by counties for chickens and chicken egg production, by

number of chickens on hand; and by counties for farms reporting chickens and turkeys by numbers raised. Washington, 1942. 739 p.

Assembles almost all data about poultry compiled by the Bureau since the First Census of Agriculture in 1840. Kept up-to-date since by the quinquennial agricultural censuses.

Miscellaneous

C214 INTERNATIONAL INSTITUTE of AGRICULTURE. L'aviculture dans le monde. Rome, 1933. 3 v.

A country-by-country report on the status of research in poultry husbandry, with extensive bibliographic citations and a list of poultry journals.

C215 NATIONAL RESEARCH COUNCIL. COMMITTEE on ANIMAL NUTRITION. Recommended nutrient allowances for domestic animals. Number 1. Nutrient requirements for poultry. Rev. (N.R.C. Pub. 301.) Washington, 1954. 27 p.

One of a frequently revised series of reference manuals on nutrition.

VETERINARY MEDICINE

Veterinary medicine, the science of the prevention, diagnosis, and treatment of diseases and injuries of animals, parallels the science of human medicine in practically every respect. Not only do both sciences concern themselves with the same topics, such as anatomy, embryology, bacteriology, hygiene, physiology, pharmacology, parasitology, obstetrics, radiology, diagnosis, surgery, and clinical techniques, but the laboratory and clinical relationships, too, are close. The experience gained in working with animals has resulted in major contributions to human medicine, and the veterinarian relies heavily on the techniques and information sources of the human specialist.

It is assumed that bibliographical work in veterinary medicine will require the use of medical reference guides of various kinds, such as the Quarterly cumulative index medicus and the Current list of medical literature. With the exception of Doe (C216) and Postell (C217) only those literature and information guides specifically devoted to veterinary medicine are described here.

Bibliographies of Bibliographies - General Medicine

A discussion of the literature and sources of information on veterinary medicine would not be complete without considering those for human medicine. The reference books in the medical field have been described in several guides, among which are the following.

C216 DOE, J. and MARSHALL, M. L. Handbook of medical library practice, with a bibliography of reference works and histories in medicine and the allied sciences. Ed. 2. Chicago, American Library Association, 1956. 630 p.

C217 POSTELL, W. D. An introduction to medical bibliography. New
Orleans, 1951. 101 p.

Abstracting Journals

As in certain other fields, the most comprehensive and useful bibliographical guides are of British origin. The abstracting journal, Veterinary bulletin (C218), and its companion bibliography, Index veterinarius (C221), provide a key to all types of world literature in the field. Chief shortcomings of these journals are delays in publication of abstracts and indexes for the Veterinary bulletin and past irregularities in meeting Index veterinarius publishing schedules. There is no comparable service in the United States. Biological abstracts (A11) covers the literature of fundamental veterinary research and abstracts approximately 1,000 items annually in this field. The Bibliography of agriculture (A24) indexes both fundamental research and the literature of applied veterinary medicine, listing in the vicinity of 6,000 or more journal articles and books a year in recent years. As compared to these, the Veterinary bulletin publishes 3,000 or more abstracts annually, and the Index veterinarius cites about 10,000 references of all types.

The Experiment station record (A22), during its years of publication, abstracted a great deal of research literature of veterinary medicine and is particularly useful for the years from 1889-1930, before the British journals began publication. Literature on nutrition and nutritional diseases is abstracted in the Nutrition abstracts and reviews (E13). Chemical aspects of nutritional diseases, pharmacology, etc., come within the scope of Chemical abstracts (D8). The latter now covers the literature of pharmaceuticals, formerly abstracted in Pharmaceutical abstracts (v. 1-12, 1935-1947). The Agricultural index (A23) lists articles published in a selected group of journals issued in the United States, as well as covering the releases of the U.S. Department of Agriculture and its agencies and the experiment stations and extension agencies.

A great many journals publish abstracts or citations to veterinary literature. The majority of these are intended to keep the reader aware of current selected literature in his field. Lack of cumulative indexes and systematic arrangement greatly reduces the reference usefulness of these journals. One of the most extensive services is provided by the journal, Veterinärmedizin, which abstracts world literature. This is the latest title in a series which has included the Jahresbericht über die Leistungen auf dem Gebiete der Veterinär-medizin (1881-1917), Ellenberger-Schütz Jahresbericht (1918-1927), and Jahresbericht Veterinär-medizin (1928-1943).

C218 COMMONWEALTH BUREAU of ANIMAL HEALTH. Veterinary bulletin. 1- April 1931- Weybridge, Eng., 1931-
Issued monthly. Superseded the Tropical veterinary bulletin (v. 1-18, no. 4, 1912-1930), which had narrower coverage of the literature. Author index in each issue, annual author and subject indexes. Abstracts world research literature, excluding publications that are popular or merely informative.

C219 COMMONWEALTH INSTITUTE of ENTOMOLOGY. Review of applied entomology. Series B: Medical and veterinary. 1- 1913-
London, 1913-

Issued monthly. Author index in each issue, separate annual author and subject indexes. Abstracts in detail world literature concerning the relations of animals and insects. See C390 for a list of the journals that are abstracted.

C220 LANDWIRTSCHAFTLICHES zentralblatt. Abt. IV: Veterinär-medizin. v. 1- December 1955- Berlin, Deutsche Akademie der Landwirtschaft, 1955-
Issued monthly or oftener. Abstracts, under broad headings, world literature on veterinary medicine. Author index in each issue.

Bibliographies and Indexes - General

C221 COMMONWEALTH BUREAU of ANIMAL HEALTH. Index veterinarius. 1- April 1933- Weybridge, Eng., 1933-
Issued quarterly. Bibliographical citations only, no abstracts. Author and subject entries for each publication alphabetized together in one sequence. Includes everything abstracted in the Veterinary bulletin (C218) and, in addition, cites informative and popular books and articles. Refers back to abstracts in the Veterinary bulletin if such are available.

C222 U.S. SURGEON GENERAL'S OFFICE. Books on veterinary medicine and related topics published during the years 1936 to 1946. Washington?, 1946. 36 p.
----- Supplement no. 1- January 1, 1948-

Bibliographies and Indexes - Early

The following are of interest as antiquarian and historical lists.

C223 BRÜCKMANN, F. E. Bibliotheca animalis oder verzeichnis der meisten schriften so von thieren und deren theilen handeln. Wolfenbüttel, 1743. 277 p.
----- Continuatio. 1747. 178 p.

C224 HENZE, J. K. G. Entwurf eines verzeichnisses veteranischer schriften und einzelner abhandlungen. Stendal, 1781. 246 p.

Bibliographies and Indexes - Library Catalogs

Of the many printed library catalogs, the following are significant.

C225 COPENHAGEN. K. VETERINAER- og LANDBOHØJSKOLE. BIBLIOTEK. Katalog over den Kongelige veterinaer- og landbohøjskoles bibliotek, indtil udgangen af 1894. København, Bang, 1898. 888 p.
----- Tillaeg omfattende bibliotekets tilvaekst, 1895-1916. København, Bang, 1921. 1,721 p.

C226 HANOVER (CITY). TIERÄRZTLICHE HOCHSCHULE. BIBLIOTEK. Katalog. Hannover, Hildebrandt, 1904. 236 p.

C227 LONDON. ROYAL COLLEGE of VETERINARY SURGEONS. LIBRARY. Catalogue of the historical collection: books published before

1850. London, 1953. 36 p.
C228 LYONS. ÉCOLE NATIONAL VÉTÉRINARIE. BIBLIOTHÈQUE. Catalogue de la bibliothèque. Lyon, Rey, 1937. 166 p.
-----Supplément 1, 1937-1944. Lyon, 1945.
C229 MILAN. R. SCUOLA SUPERIORE di MEDICINA VETERINARIA. BIBLIOTECA. Catalogo. Milano, Premiata Tipografia "Agraria," 1908. 311 p.
C230 STOCKHOLM. VETERINÄRHÖGSKOLAN. Katalog. Stockholm, 1936. 583 p.
C231 UTRECHT. RIJKSUNIVERSITEIT. VEEARTSENIJKUNDIGE FACULTEIT. BIBLIOTHEEK. Catalogus van de Veeartsenijkundige faculteit. Utrecht, 1926-1928. 973 p.

Bibliographies and Indexes - Regional

C232 CONGRESO PANAMERICANO de MEDICINA VETERINARIA. I, Lima, 1951. Contribución a la bibliografia veterinaria americana, recopilada por R. Ponce Paz. Lima, 1951. 47 p.
Includes a list of theses in veterinary medicine presented at selected Latin-American universities.
C233 KLEE, R. Bibliotheca veterinaria, oder verzeichnis sämtlicher bis zur gegenwart im deutschen buchhandel erschienenen bücher und zeitschriften auf dem gebiete der veterinärwissenschaften. Leipzig, Seemann, 1901. 247 p.
Presumably includes publications listed in earlier bibliographies of German veterinary literature such as Enslin (1843), Büchting (1867), Gracklauer (1883), etc.
C234 LISBON. ESCOLA SUPERIOR de MEDICINA VETERINARIA. Indice bibliográfico dos escritos produzidos pelos autores veterinàrios portugueses. Lisbon, Silvas, 1936. 229 p.
C235 VERGANI, F. and DIAZ UNGRIA, C. Bibliografia veterinaria venezolana (fichero venezolana de bibliografia veterinaria y ganadera). Venezuela. In Inst. de Invest. Vet. B. 5(1953):715-1869.

Bibliographies and Indexes - Veterinary Zoology and Entomology

C236 FRANCE. OFFICE de la RECHERCHE SCIENTIFIQUE et TECHNIQUE OUTRE-MER. Bulletin signaletique d'entomologie médicale et vétérinaire. 1-   1953-   Paris, 1953-
A quarterly bulletin containing titles of articles on medical and veterinary entomology appearing in over 200 periodicals published or received in France during the preceding quarter. In the first part, articles are arranged under subject headings in such a way that they can be cut out to form index cards: names of new species of insects are indicated when they are not mentioned in the title. In the second part, periodicals scrutinized are arranged alphabetically and articles are listed in the order in which they appear in the periodical.
C237 U.S. BUREAU of ANIMAL INDUSTRY. Index-catalogue of medical and veterinary zoology. Washington, 1932-1952. 5711 p. 18 pts. in 7 v.

-----Supplement 1- Washington, 1953-
Incorporates and brings up to date a previous publication of the same title. Most comprehensive for literature on cestodes, nematodes, trematodes, and thornheaded worms. Citations arranged alphabetically by author. No comprehensive subject index as yet.

Bibliographies and Indexes - Special

Bibliographies on specific topics can be located by means of the subject indexes of the Bibliography of agriculture (A24), the Experiment station record (A22), Index veterinarius (C221), Bibliographic index (A2), etc. Following are a few of these topical bibliographies:

C238 BIBLIOGRAPHY of poultry diseases. v. 1-10, no. 1, April 1936-April 1945. New Brunswick, New Jersey Agricultural Experiment Station, 1936-1945.
Issued semi-annually during the years it was published, covering virus diseases, bacterial diseases, protozoan diseases, toxicology, etc.
C239 DIECKERHOFF, W. Geschichte der rinderpest und ihrer literatur. Berlin, 1890. 270 p.
C240 HEKMEYER, F. C. Korte geschiedenis der runderpest, benevens eene opgave van al de over deze ziekte handelende geschriften, die van de vroegste tijden tot op heden zijn uitgekommn. Amersfoort, Meijers, 1845. 82 p.
C241 MERCK & CO., INC. Annotated bibliography, penicillin in veterinary medicine. Rahway, N. J., 1950. 50 p.
C242 MERCK & CO., INC. Streptomycin and dihydrostreptomycin in veterinary medicine; annotated bibliography. Rahway, N. J., 1950. 15 p.

Encyclopedias and Handbooks

Of the many encyclopedic works on veterinary medicine that have been published, particularly in Germany, France, and England, only a few are currently up to date and in print.

C243 BAKER, E. T. The home veterinarian's handbook, a guide for handling emergencies in farm animals and poultry. New York, Macmillan, 1943. 191 p.
Relatively nontechnical. Contains about 700 articles in alphabetical order on diseases and treatments.
C244 SEIDEN, R. Livestock health encyclopedia. New York, Springer, 1951. 614 p.
C245 UEBELE, G. Handlexicon der tierärztlichen praxis. Ed. 7. Stuttgart, Medica, 1953-1954. 2 v.
C246 U.S. BUREAU of ANIMAL INDUSTRY. Special report on diseases of cattle. Rev. ed. Washington, 1942. 507 p.
C247 U.S. BUREAU of ANIMAL INDUSTRY. Special report on diseases of the horse. Rev. ed. Washington, 1942. 584 p.
C248 U.S. DEPARTMENT of AGRICULTURE. Keeping livestock healthy. (Ybk. of Agr. 1942) 1942. 1276 p.

A popular guide to diseases and parasites that attack domestic animals. For each disease, history, symptoms, diagnosis, treatment, method of eradication, and literature references are given.

Dictionaries

Because of the similarity of terminology in human and veterinary medicine, standard medical dictionaries form an important part of the veterinary reference collection. These can be found listed in Winchell (A8) and in the Handbook of medical library practice (C216). Some of the encyclopedic works in the field define terms in considerable detail: Seiden's Livestock health encyclopedia (C244), Seiden's Poultry handbook (C187), etc.

C249 AMERICAN VETERINARY MEDICAL ASSOCIATION. COMMITTEE on REVISION of VETERINARY ANATOMICAL NOMENCLATURE. Nomina anatomica veterinaria. Detroit, 1923. 60 p.
    An attempt to standardize terminology.

C250 CAGNY, P. Dizionario veterinario. Torino, Unione Tipografico - editrice Torinese, 1907-1910. 2 v.

C251 CAGNY, P. and COBERT, H. J. Dictionnaire vétérinaire. Paris, Baillière, 1904. 2 v.

C252 COBACHO, J. G. Diccionario de términos técnicos de veterinaria. Madrid, Cosano, 1926. 684 p.

C253 FONTAINE, A. Nouveau dictionnaire vétérinaire, médecine, chirurgie, thérapeutique, législation sanitaire et sciences qui s'y rapportent. Paris, Baillière, 1921-1924. 2 v.

C254 KITTAKU, R. Moderne tierärztliche terminologie. Tokyo, Buneido, 1942. 404 p.
    Polyglot, including Japanese.

C255 KRÜGER, G. Der anatomische wortschatz unter mitberücksichtigung der histologie und der embryologie für die studierenden der veterinärmedizin. Ed. 3. Leipzig, Hirzel, 1953. 104 p.

C256 LITTRÉ, É. Dictionnaire de médecine, de chirurgie, de pharmacie, de l'art vétérinaire et des sciences qui s'y rapportent. Éd. 21. Paris, Baillière, 1908. 1,842 p.
    Lists equivalent terms in French, Greek, Latin, German, English, Italian, and Spanish.

C257 MILLER, W. C. and WEST, G. P. Black's veterinary dictionary. Ed. 3. London, Black, 1953. 1,112 p.
    The most extensive English-language veterinary dictionary.

C258 PETIT dictionnaire vétérinaire. Ed. 2. Paris, n. d. 146 p.

C259 PRETORIA. VETERINARY RESEARCH INSTITUTE. Draft of an international list of animal diseases, comp. by the veterinary staff of the Veterinary Research Institute. Pretoria, 1938. 80 p.
    Mostly English terms, with some French and German equivalents.

C260 SEWELL, A. J. The dog's medical dictionary. New York, Philosophical Library, 1952. 271 p.

C261 VETERINARNYĬ entsiklopedicheskiĭ slovar'. Moscow, Gos. Izd-vo Selkhoz. Lit-ri, 1950. 2 v.

C262 WHITE, J. A compendious dictionary of the veterinary art. London, Longman, 1817. 344 p.
    An example of an early veterinary dictionary.

C263 WIRTH, D. Lexikon der praktischen therapie und prophylaxie für thierärzte. Wien, Urban & Schwarzenberg, 1948-1949. 2 v.

Directories, Biography

The following are representative of the available national and local directories of veterinarians. A particularly convenient index to directories is provided by the Bibliography of agriculture (A24), which, in its annual subject index, lists directories under the heading "Veterinarians - Directories." The recently established series, Working references of livestock regulatory establishments (C98), includes veterinarians and disease-control officials of the federal government and local governments and supersedes a number of lists of these officials published in the past. (See note on directories, p. 15).

C264 AMERICAN VETERINARY MEDICAL ASSOCIATION. Directory. 1923/24-   Chicago, 1924-
  Issued biennially. 1954 edition includes United States and Canada, lists veterinary colleges of the world, veterinary hospitals in the United States and Canada, veterinary medical associations, accredited schools, journals, commercial firms, some history of schools, and members of the Association.

C265 ANNUAIRE des docteurs en médecine vétérinaire de Belgique. Ed. 18. 1953. Bruxelles, 1953. 74 p.

C266 ANNUAIRE vétérinaire. Paris, 1921-
  Lists veterinarians, veterinary schools, etc., in France. Suspended 1940-1947.

C267 LONDON. ROYAL COLLEGE of VETERINARY SURGEONS. A register of members of the Royal College of Veterinary Surgeons, from January 1794 to December 1851, inclusive. London, Crompton, 1852. 41 p.

C268 LONDON. ROYAL COLLEGE of VETERINARY MEDICINE. The register of veterinary surgeons and the supplementary veterinary register. 1953. London, 1953. 482 p.
  Issued periodically. Continues C267.

C269 NEUMANN, L. G. Biographies vétérinaires. Paris, Asselin, 1896. 443 p.

C270 SCHRADER, G. W. Biographisch-literarisches lexicon der thierärzte aller zeiten und länder sowie der naturforscher, aertze, landwirthe, stallmeister, u.s.w., welche sich um die thierheilkunde verdient gemacht haben. Stuttgart, Ebner, 1863. 490 p.

C271 WISCONSIN. DEPARTMENT of AGRICULTURE. Wisconsin registered licensed veterinarians. (Wis. Dept. Agr. B. 323) 1954. 29 p.
  Many of the states publish similar directories.

History

A great many histories of veterinary medicine and of specialized subjects within the field are available. Only representative titles are listed here.

C272 BIERER, B. W. American veterinary history. Parts I-VII. Baltimore, 1940. 222 p.

C273 BIERER, B. W. History of animal plagues of North America. Baltimore, 1939. 114 p.
C274 BIERER, B. W. A short history of veterinary medicine in America. East Lansing, Mich., Michigan State University Press, 1955. 124 p.
C275 DANSK veterinaerhistorisk aarbog. Aarg. 1- 1934- Skive, 1935-
C276 FLEMING, G. Animal plagues; their history, nature, and prevention. London, Chapman, 1871-1882. 2 v.
C277 FONTAINE, H. Das deutsche heeresveterinärwesen. Seine geschichte bis zum jahre 1933. Hanover, Schaper, 19(?). 1,360 p.
C278 FROEHNER, R. Kulturgeschichte der tierheilkunde. v. 1- Konstanz, Terra-Verlag, 1952-
   To be completed in three volumes: I, Antiquity; II, History of veterinary medicine in Germany; and III, History of veterinary medicine in the rest of the world. Bibliographic citations are noteworthy.
   V. I (1952) and V. II (1954) issued to date.
C279 HANSEN, J. J. P. Den Danske dyrlaegeforening, 1849-1949. Danske dyrlaeger, 1779-1947. København, Mortensen, 1949. 545 p.
C280 LAIGNEL-LAVASTINE, M. Histoire générale de la médecine, de la pharmacie, de l'art dentaire et de l'art vétérinaire. Paris, Michel, 1936-1949. 3 v.
C281 LECLAINCHE, E. Histoire de la médecine vétérinaire. Toulouse, Office du Livre, 1936. 812 p.
C282 MERILLAT, L. A. and CAMPBELL, D. M. Veterinary military history of the United States. Kansas City, Mo., Haver-Glover Laboratory, 1935. 2 v.
C283 SANZ EGAÑA, C. Historia de la veterinaria española; albeitería, mariscalería, veterinaria. Madrid, Espasa-Calpe, 1941. 493 p.
C284 SCHMALTZ, R. Entwicklungsgeschichte des tierärztlichen berufes und standes in Deutschland. Berlin, Schoetz, 1936. 490 p.
C285 SENET, A. Histoire de la médecine vétérinaire. Paris, Presses Universitaire de France, 1953. 114 p.
C286 SHIRAI, T. History of Japanese veterinary medicine. Tokyo, Buneido, 1944. 716 p.
   In Japanese.
C287 SMITH, F. The early history of veterinary literature and its British development. London, Baillière, 1919-1933. 4 v.

Periodical Lists

The most complete lists of veterinary serials and of journals in allied fields are in the Veterinary bulletin (C218) (see v. 25(1955):i-xii) and Index veterinarius (C221) (see v. 22(1955):ix-xx). Veterinary journals are also listed in the following.

C288 PAN AMERICAN SANITARY BUREAU. List of medical and public health journals of Latin America. (P. no. 223.) Washington, 1946. 43 p.
C289 WORLD HEALTH ORGANIZATION. World medical periodicals. Les périodiques médicaux dans le monde. Periódicos médicos del mundo. Paris, 1953. 237 p.

ECONOMIC ENTOMOLOGY

Economic entomology is concerned with the relations of insects to crops and domestic animals and with insects as productive, economic factors. The economic entomologist seeks to control injurious insects. For this he needs information on the life history, distribution, classification and identification, physiology, anatomy, ecology, and biological and chemical control of insects. For economically productive insects, such as bees, he is concerned with culture, reproduction, diseases, and other factors involved in protection and improvement. Medical entomology, so far as it concerns insects that annoy, infest, or otherwise affect domestic animals, is also of interest to the economic or agricultural entomologist. So is the study of insects as indirect factors, or vectors, of plant and animal disease.

The total literature on insects comprehends millions of books, journal articles, and documents, and is increasing at a prodigious rate; nearly 2,000 serial publications contain material of interest to entomologists. Bibliographical control of such a field, therefore, presents great difficulties.

The attempt to divide the field into such subjects as medical entomology, economic entomology, and systematic entomology, creates artificial literature boundaries that cannot be maintained in actual practice. However, in this section the literature of economic entomology is emphasized, and that of medical entomology is listed only so far as it deals with domestic animals. Reference is made to the literature of broader fields when it seems desirable. The literature of apiculture and of chemical insect control (insecticides) is discussed in separate sections.

Bibliographies of Bibliographies

An extensive class of reference literature consists of catalogs, keys, and bibliographical guides to orders, suborders, genera, and species of insects. Because there is so much of it, this class of literature is excluded from the following list. Outstanding examples will be found cited in Brues (C394), Chamberlin (C290), and most entomological textbooks. The Zoological record (C10) is a particularly useful current guide to this type of systematic literature. Smith's Guide to the literature of the zoological sciences (C1), the work of an entomologist, describes the bibliographical resources of zoology, with some emphasis on the relation of entomology to the broader field.

C290 CHAMBERLIN, W. J. Entomological nomenclature and literature. Ed. 3. Dubuque, Ia., Brown, 1952. 141 p.
    Compiled by a professional entomologist, this guide describes the bibliographical resources in the field of zoology, with particular emphasis on entomology. It gives extensive coverage of entomological bibliography, and is intended for the entomologist conducting a professional literature search. Contains descriptive information on government documents and annotated lists of journals.

C291 SHERMAN, J. D. Some entomological and other bibliographies. In N. Y. Ent. Soc. J. 32(1924):206-215.
    This survey, by a bookman in the field of entomology, has continuing interest because of its historical sidelights.

Abstracting Journals

Economic entomology has a specialized abstracting service, the British Review of applied entomology (C219 and C293), whose most serious defects are the delayed subject indexes and the lack of systematic arrangement in the monthly issues. Biological abstracts (A11) is widely used by economic entomologists (in 1953, it published over 700 abstracts of literature of interest to entomologists). It overlaps the Review of applied entomology, but each covers some literature not abstracted by the other, and both may abstract the same literature at different times, varying by as much as a year or more. Abstracts in the British journal are, in general, much longer.

Literature on pesticides and insect-control chemicals are abstracted in Chemical abstracts (D8), as well as a considerable amount of literature on chemical aspects of insect metabolism, physiology, and chemical relations of insects and plants. Additional abstracts, particularly on control, will be found in Field crop abstracts (B84), Herbage abstracts (B85), Horticultural abstracts (B83), and Plant breeding abstracts (B129). Agricultural and horticultural engineering abstracts (D55) covers world literature on mechanical devices for insect control. Among discontinued abstract services, the Experiment station record (A22) is still of great potential value because of its long abstracts of literature published as far back as the late 1880's. Economic entomology is represented by the section "Economic zoology - Entomology." The "Reviews and abstracts" section of the journal Mosquito news selectively abstracts world literature on mosquito control and malariology.

C292 ACRIDOLOGICAL abstracts. 1- 1950- London, Anti-Locust Research Centre (British Museum Nat. Hist.), 1950-
Issued irregularly. Abstracts world locust-control literature.

C293 COMMONWEALTH INSTITUTE of ENTOMOLOGY. Review of applied entomology. Series A: Agricultural. 1- 1913- London, 1913-
Issued monthly. Author index in each issue; separate annual author and subject indexes which are late in appearing. Abstracts are long and detailed. Over 750 publications abstracted in 1952, including not only entomological journals, but those from other fields that publish literature of interest to entomologists. See C390 for a list of the journals abstracted. This is the basic abstracting journal for economic entomologists, especially when used in conjunction with its companion, Series B: Medical and veterinary (C219).

Bibliographies and Indexes - General

Probably the most extensive listing of entomological publications, particularly those stressing the literature of insect control and agricultural entomology, is to be found in the Bibliography of agriculture (A24). During 1953, more than 10,000 separate publications were cited under the heading "Entomology." This probably included most of the literature that would subsequently be abstracted in either Biological abstracts or the Review of applied entomology and in addition a great deal more. The Bibliography of agriculture, therefore, is first choice where a title entry is sufficient. The Agricultural index (A23) is particularly to be recommended for its listing of selected literature on economic entomology published in the United States.

One of the most widely used indexing services in the field of entomology is the Zoological record (C10). By editorial policy, the literature of economic entomology is covered by the Review of applied entomology and not by the Zoological record, which cites the literature of the fundamental science, systematic literature, etc. The distinction is a difficult one to maintain, and the Zoological record is a basic publication in any entomological library.

A great many current journals have bibliographical sections of varying scope. Because of broad subject arrangement, lack of indexing, etc., these are chiefly useful as current guides to publishing and have little practical value for retrospective literature searching. One of the most comprehensive is that published in the Entomological news, a bibliographical list of articles appearing in over one hundred journals received at the Academy of Natural Sciences, Philadelphia, and the University of Pennsylvania. The citations are arranged under broad headings or under specific, systematic names. January and June issues contain lists of journals that are scanned. Mosquito news has a "Bibliography" section that covers world literature. The Bulletin signaletique d'entomologie medicale et veterinaire (C236) is also useful.

To summarize: for the most exhaustive guide to current literature of economic entomology, the Bibliography of agriculture; for the best coverage by abstracts of the field, the Review of applied entomology supplemented by Biological abstracts; for the best citation list to entomological literature other than economic, Zoological record; for the most complete citation list of North American economic entomology, the Index to the literature of American economic entomology (C311); for older literature the Bericht über die wissenschaftlichen Leistungen im Gebiete der Entomologie (C294) since 1834, and the Zoological record, since 1863; and for abstracts, the Experiment station record from 1889 to 1946, the Review of applied entomology since 1913, and Biological abstracts since 1926.

C294 BERICHT über die wissenschaftlichen leistungen im gebiete der entomologie. 1834-1917. Berlin, 1835-1918.
    A republication of the entomological sections of the Archiv für Naturgeschichte. Abt. B, Jahresberichte. These were reviews and bibliographies of literature. The republication was in the form of Beihefte to the Deutsche Entomologische Zeitschrift.

C295 ENTOMOLOGISCHE literaturblätter. Repertorium der neuesten arbeiten auf dem gesammtgebiet der entomologie. v. 1-14, 1901-1914. Berlin, 1901-1914.
    Supplement to the Deutsche Entomologische Zeitschrift. Appeared as Repertorium Entomologicum (C296) from 1924 to 1933. Lists entomological contents of journals published throughout the world, with an added list of books and articles by broad subjects.

C296 REPERTORIUM entomologicum. v. 1-10, 1924-1933. Berlin, 1924-1935.
    Supplement to Deutsche Entomologische Zeitschrift. Appeared as Entomologische Literaturblätter (C295) from 1901-1914. Lists entomological contents of journals.

C297 U.S. BUREAU of ENTOMOLOGY and PLANT QUARANTINE. LIBRARY. Entomology current literature. v. 1-11, 1932-1942.

Washington, 1932-1942. 11 v.
This was superseded by corresponding sections of the Bibliography of agriculture (A24).

Bibliographies and Indexes - Early

Literature concerned with the control of insects affecting man's economic life was not listed separately to any significant degree before the first decade of the nineteenth century. For the broad field of entomology, however, there is a virtually continuous sequence of comprehensive bibliographies covering world literature from earliest times: Horn and Schenkling (C300), the Zoological record (C10), etc. Bibliographies published before Horn and Schenkling are cited below for historical reasons only; practically all verifiable references in these earlier works were incorporated into the Index litteraturae entomologicae (C300).

Entomological literature was included in the various zoological bibliographies of the nineteenth century: the Zoological record, with its emphasis on systematic entomology; Engelmann's Bibliotheca historico-naturalis (C14), including literature published from 1700-1846, and its successor, the Bibliotheca zoologica (C12, C13), which carried the indexing to 1880. The Zoologischer Anzeiger covered the literature from 1881-1895, at which time the citation list began to be put out independently as the Bibliographia zoologica (C6), which continued to 1934.

C298 EISELT, J. N. Geschichte, systematik und literatur der insectenkunde, von den ältesten zeiten bis auf die gegenwart. Leipzig, Hartmann, 1836. 255 p.
Extensive bibliography of early literature, list of entomological bibliographies available at the time, and a history of entomology.

C299 HAGEN, H. A. Bibliotheca entomologica. Die litteratur über das ganze gebiet der entomologie, bis zum jahre 1862. Leipzig, Engelmann, 1862-1863. 2 v.
Lists of bibliographies, histories, biographies, entomological societies and their publications, etc. Has a subject index, which must be used as the only available subject index to Horn and Schenkling (C300). Continued to some extent for the period 1861-1880 by the Bibliotheca zoologica II (C13), but Horn and Schenkling provides the definitive continuation.

C300 HORN, W. and SCHENKLING, S. Index litteraturae entomologicae. Ser. 1: Die welt-literatur über die gesamte entomologie bis inklusive 1863. Berlin-Dahlem, 1928-1929. 4 v. in 1.
A revision of Hagen and earlier bibliographies, presumably including all verifiable earlier literature citations plus several thousand new entries. Intended to cover the literature up to the first volume of the Zoological record (C10). No index, but subject and author indexes to Hagen may be used for the majority of references. Arranged alphabetically by author.

C301 NODIER, C. Bibliographie entomologique; ou, Catalogue raisonné des ouvrages relatifs à l'entomologie et aux insectes . . . Paris, Moutardier, 1801. 64 p.

C302 PERCHERON, A. R. Bibliographie entomologique. Paris, Baillière, 1837. 2 v.

World literature from the earliest times through 1834, including society publications; list of journals and societies. With indexes. For criticism, see introduction to Hagen (C299). Presumably corrected and incorporated into Hagen, later into Horn and Schenkling (C300).

Bibliographies and Indexes - Library Catalogs

Published catalogs of libraries are a dwindling genus, but those of special entomological libraries can give clues to important literature, particularly writings not generally cited in bibliographies of research papers. Appropriate sections of the Catalogue systematique (A30) of the Library, International Institute of Agriculture also cite a considerable collection of entomological literature.

C303 AMERICAN ENTOMOLOGICAL SOCIETY, Philadelphia. Catalogue of works in the library . . . With its Transactions, v. 1, 1867-1868. Philadelphia, 1868. 32 p.
C304 DEUTSCHES ENTOMOLOGISCHES MUSEUM. Katalog der bibliothek. Berlin-Dahlem, 1913. v. 1.
C305 ENTOMOLOGISCHER VEREIN, Stettin. Katalog der bibliothek. With Ent. Ztg. Stettin, 1885, Jahrg. 46. Stettin, Grassmann, 1885. 105 p.
C306 NEDERLANDSCHE ENTOMOLOGISCHE VEREENIGING. Catalogus der bibliotheek. Ed. 4. (Incl. Sup. 1, 2.) Amsterdam, 1938.
----- Supplement 3. 1939.
C307 ROYAL ENTOMOLOGICAL SOCIETY of LONDON. Catalogue of the library. London, 1893. 312 p.
----- Supplementary catalogue. London, 1900. 147 p.
C308 U.S. DEPARTMENT of AGRICULTURE. LIBRARY. Catalogue of the publications relating to entomology in the library of the U.S. Dept. of Agriculture. (Library B. 55) 1906. 562 p.

Bibliographies and Indexes - North American

C309 BANKS, N. A list of works on North American entomology. (U.S. Bur. of Ent. B. 81) 1910. 120 p.
C310 BANKS, N. A list of works on North American entomology. (U.S. Bur. of Ent. B. 24, n.s.) 1900. 95 p.
C311 INDEX to the literature of American economic entomology. v. 1-
1905/14-   Melrose Highlands, Mass., American Association of Economic Entomology, 1917-
The first volume covered literature published from 1905-1914, subsequent volumes covered five-year periods through 1945-1949, annual since 1950. Compiled in the U.S. Department of Agriculture Library. Includes literature referring to the economic entomology of the United States, Alaska, Canada, Mexico, Canal Zone, Cuba, Puerto Rico, Hawaii, and, since v. 7, certain Pacific Islands that were important in World War II. Arranged alphabetically by subject, citations indicate whether the publications have bibliographies, illustrations, keys, etc. Sometimes called the Colcord list. It was preceded by C312.

C312 U.S. BUREAU of ENTOMOLOGY. Bibliography of the more important contributions to American economic entomology. Washington, 1889-1905. 8 v.
I, The more important writings of Benjamin Dana Walsh, by Samuel Henshaw. II, The more important joint writings of B. D. Walsh and C. V. Riley, by Samuel Henshaw. III, The more important writings of Charles Valentine Riley, by Samuel Henshaw. IV, The more important writings of government and state entomologists and of other contributors to the literature of American economic entomology; A-K, by Samuel Henshaw. V, The more important writings of government and state entomologists and of other contributors to the literature of American economic entomology; L-Z, by Samuel Henshaw. VI, The more important writings published between June 30, 1888 and December 30, 1896, by Nathan Banks. VII, The more important writings published between December 31, 1896 and January 1, 1900, by Nathan Banks. VIII, The more important writings published between December 31, 1899 and January 1, 1905, by Nathan Banks. Sometimes referred to as the Henshaw and Banks list.

C313 U.S. DEPARTMENT of AGRICULTURE. DIVISION of ENTOMOLOGY. An enumeration of the published synopses, catalogues, and lists of North American insects; together with other information intended to assist the student of American entomology. (B. 19) 1888. 77 p.

Bibliographies and Indexes - Canadian and United States Government Publications

Perhaps the greatest quantity of literature on insect control is published by government departments and international agencies. The best current bibliographic coverage of this material is provided by the Bibliography of agriculture (A24), with special listings of publications of the U. S. Department of Agriculture, experiment stations, extension services, F.A.O., etc., at the back of each issue. The Agricultural index (A23) provides practically complete coverage of agricultural publications of federal governmental agencies in the United States, as well as listing many publications of states, the F.A.O., etc. The monthly catalog of United States government publications, is useful for its references to publications of interest to etomologists issued by agencies other than the U.S. Department of Agriculture. The Library of Congress' Monthly checklist of state publications may include references to releases by state insect-control agencies and other state agencies not covered in any of the other current bibliographical series.

In addition to the general indexes to federal agricultural publications listed in A35-A44, which include entomological releases, there are a few bibliographies intended as specific guides to entomological publications of the federal government.

C314 CANADA. DEPARTMENT of AGRICULTURE. ENTOMOLOGICAL BRANCH. Index to entomological publications of the Department of Agriculture, 1884-1936, by C. E. Petch. Ottawa, 1938. 410 p.

C315 COLCORD, M. Check list of publications on entomology issued by the

United States Department of Agriculture through 1927, with subject index. (U.S.D.A. Libr. Bibliog. Contrib. 20) 1930. 261 p.

C316 U.S. BUREAU of ENTOMOLOGY. DIVISION of CEREAL and FORAGE INSECT INVESTIGATIONS. List of entomological publications of personnel of Cereal and Forage Insect Investigations, U.S. Bur. of Entomology, 1904-1928, inclusive, by J. S. Wade. Washington, 1929. 46 p.

C317 U.S. BUREAU of ENTOMOLOGY and PLANT QUARANTINE. DIVISION of INSECT PEST SURVEY and INFORMATION. List of entomological technique (ET-series) publications. Washington, 1935-1951. 6 nos.

Lists nos. 1-291 of the ET series. Lists for 234-270 as ET 274, for 271-291 as ET 292; other lists unnumbered. No list issued for ET 292-314. ET 314(1954) concluded the series, which was replaced by the new ARS series.

C318 U.S. SUPERINTENDENT of DOCUMENTS. Insects: worms and insects harmful to man, animals, and plants. Ed. 46. (Price List 41) 1955. 8 p.

Revised approximately annually. Lists only in-print publications available for sale.

C319 WADE, J. S. On entomological publications of the United States government. In Ent. Soc. of Wash. Proc. 25(1923):1-32.

Summarizes earlier federal publication series of interest to the entomologist.

Bibliographies and Indexes - Other Countries

Only a few examples of the numerous entomological bibliographies relating strictly to specific countries, other than Canada and the United States, are noted here.

C320 LEONARD, M. D. Annotated bibliography of Puerto Rican entomology. In Puerto Rico. Dept. of Agr. J. 17(1933):5-96.

C321 MUSGRAVE, A. Bibliography of Australian entomology, 1775-1930, with biographical notes on authors and collectors. Sidney, Royal Zoological Society of New South Wales, 1932. 380 p.

C322 OUCHI, Y. Bibliographical introduction to the study of Chinese insects. (Shanghai Sci. Inst. J., Sect. III, Vol. 2) Shanghai, Kelly, 1934. 533 p.

C323 PFEFFER, A. Bibliographie entomologique de la Bohême, de la Moravie, et de la Silésie pour les années 1800-1940. In Českoslov. Společnost. Ent. Časopis 47(1950):231-356.

C324 PIRÁN, A. A. Bibliografía entomológica Argentina. (Argentina. Inst. de Sanid. Veg. P. Ser. B., v. 2, no. 5) Buenos Aires, 1946. 144 p.

C325 PIRÁN, A. A. Bibliografía entomológica Argentina, 1874-1952. Buenos Aires, Museo Argentina de Ciencia Naturalista "Bernardino Rividavia." (Soc. Ent. Argentina. Curso de Ent. X-   .) 1954-

In process. First volume covers A-H.

Economic Entomology

Bibliographies and Indexes - Insect Control

There are innumerable bibliographical compilations on various phases of insect control, of which those listed here are examples.

C326 IMPERIAL BUREAU of PLANT BREEDING and GENETICS, Cambridge, Eng. Bibliography on insect pest resistance in plants (with a supplement on resistance to nematodes). Cambridge, Eng., School of Agriculture, 1944. 39 p.
C327 KIRKALDY, G. W. A bibliography of sugar-cane entomology. (Hawaiian Sugar Planters' Assoc. Expt. Sta. Div. of Ent. Rpt. of Work B. no. 8.) Honolulu, 1909. 73 p.
C328 MASERA, E. La malattie infettive degli insetti e loro indice bibliografico. Bologna, Cappelli, 1936. 343 p.
C329 OUDEMANS, A. C. Kritisch historisch overzicht der acaralogie. Leiden, Brill, 1926-1937. 9 v. in 8.
C330 WADE, J. S. A selected bibliography of the insects of the world associated with sugar cane, their predators and parasites. (Internatl. Soc. of Sugar Cane Technol. Mem. no. 1.) Honolulu, 1951. 113 p.

Dictionaries

An entomological collection will require general scientific dictionaries, chemical dictionaries, medical dictionaries, and glossaries and dictionaries in various biological fields. These can be found listed and evaluated in bibliographical guides to particular subjects, such as chemistry, and in Winchell (A8). UNESCO's Bibliography of interlingual, scientific, and technical dictionaries (A53) is useful in making selections in this specialized field. Reference can also be made to the dictionaries described in the section on economic zoology.

Entomological dictionaries were published at least as early as the eighteenth century. The following list is indicative rather than inclusive.

C331 BROOKLYN ENTOMOLOGICAL SOCIETY. Explanation of terms used in entomology. Brooklyn, 1883. 37 p.
C332 DE LA TORRE-BUENO, J. R. A glossary of entomology; Smith's "An explanation of terms used in entomology," completely rev. and rewritten. Lancaster, Pa., Scientific Press, 1937. 336 p.
Defines over 8,000 terms in entomology and such related sciences as embryology, cytology, physiology, morphology, genetics, ecology, etc.
C333 GAUL, A. T. A glossary of terms and phrases used in the study of the social insects. In Ent. Soc. Amer. Ann. 44(1951):473-484.
C334 GHIDINI, G. M. Glossario di entomologia. Brescia, "La Scuola," 1949. 260 p.
C335 HELFER, J. W. Terminologia entomologica. Ticini Regii, Bizzoni, 1832. 40 p.
C336 HERING, M. Lepidopterologisches wörterbuch. Stuttgart, Kernen, 1940. 123 p.

C337 ISHIDA, M. The dictionary of terms used in entomology (English-Japanese). Sapporo, Japan, Ishida, 1933. 294 p.
C338 JARDINE, N. K. The dictionary of entomology. London, Janson, 1931. 259 p.
Includes terms from allied sciences.
C339 KARNY, H. Der insektenkörper und seine terminologie. Wien, Pichler, 1921. 92 p.
C340 MÜLLER, C. H. Lexicon entomologicum, oder entomologisches wörterbuch. Erfurt, Henser, 1795. 704 p.
C341 MÜLLER, J. Terminologia entomologica. Ed. 2. Brünn, Winiker, 1872. 306 p.
C342 SAY, T. A glossary to Say's Entomology. Philadelphia, Mitchell, 1825. 37 p.
C343 SMITH, J. B. Explanation of terms used in entomology. Brooklyn, Brooklyn Entomological Society, 1906. 154 p.

Nomenclatural Guides

New insect names are listed in the Zoological record (C10), which since 1945, has kept Neave's Nomenclator zoologicus (C38) up-to-date. Chamberlin (C290) and Smith (C1) describe in some detail nomenclatural literature and problems.

The following nomenclatural guides are supplemented and reëdited from time to time:

C344 AMERICAN ASSOCIATION of ECONOMIC ENTOMOLOGISTS. Common names of insects approved by the American Association of Economic Entomologists. Concord, N. Y., 1950. In J. Econ. Ent. 43(1950):117-138.
C345 ROYAL ENTOMOLOGICAL SOCIETY of LONDON. COMMITTEE on GENERIC NOMENCLATURE. Generic names of British insects. Pt. 1- London, 1934-

Directories

Amateur and professional collectors and systematic students of insects are to be found in all countries. Directories have been available since at least the beginning of the nineteenth century, as the following list shows. Most directories also give information on societies and journals, occasionally on reference publications, on museums and museum collections, and on private collections.

For background information and bibliographical references on American entomological societies before 1865, Max Meisel's Bibliography of American natural history (B11) (Brooklyn, Premier, 1924-1929; 3 v.) is useful. The latest edition of Entoma (C444) contains information on entomological societies in the United States and Canada, and also provides information on suppliers of entomological equipment, consulting entomologists, laboratories, book dealers, and personnel of government agencies. (See note on directories, p. 15.)

C346 HALL, D. G. How to get further information on insects. In U.S.D.A. Ybk. of Agr. 1952, Insects, p. 737-743.

C347 AMERICAN ASSOCIATION of ECONOMIC ENTOMOLOGISTS. Membership list. In J. Econ. Ent. 45(1952):167-200.
C348 ENTOMOLOGICAL SOCIETY of AMERICA. List of members. In Ent. Soc. Amer. Ann. 45(1952):144-173.
C349 GISTEL, J. Die jetzt lebenden entomologen, kerffreunde und kerfsammler Europa's und deren übrigen continente. Munich, 1834. 80 p.
   This appeared in a new edition of 103 p. in 1836, a copy of which was not available for examination.
C350 GISTEL, J. Lexikon der entomologischen welt, der carcinologischen und arachnologischen. Adressenbuch der lebenden entomologen und entomophilen etc.; der carcinologen und arachnologen sammt ihren schriften, dann der naturforscher-akademien und deren verhandlungen, der zoologischen ephemeriden, bibliographien, biographien und real-wörterbücher, der öffentlichen und privat-sammlungen der welt, der schriften über sammlungs- und aufbewahrungsweise der gliederthiere, mit doppelten registern und einer äufzahlung aller entomologischen, carcinologischen und arachnologischen schriftsteller von Aristoteles an bis zur gegenwart. Stuttgart, Schweizerbart, 1846. 326 p.
C351 GISTEL, J. N. Die naturforscher diess- und jenseits der oceane. Reise- und correspondenz-handbuch für geologen, geognosten und mineralogen, botaniker, zoologen ganz besonders, entomologen, anatomen, rationelle aertze, astronomen, physiker, chemiker und pharmazeuten. Straubing, Schorner, 1856. 372 p.
C352 HOFFMANN, A. Entomologen-adressbuch; annuaire des entomologistes; entomologist's directory. Vienna, 1930. 368 p.
C353 JUNK, W. Entomologen-adressbuch. The entomologist's directory. Annuaire des entomologistes. Berlin, 1905. 296 p.
C354 A LIST of the existing entomological societies in the United States and Canada. In Ent. News 42(1931):126-130.
C355 NATURALIST's directory. Ed. 1, 1877-   Salem, Mass., Cassino, 1877-
   Gives names, addresses, special subject interest of amateur and professional entomologists from most parts of the world; list of natural history periodicals, museums.
C356 ROYAL ENTOMOLOGICAL SOCIETY of LONDON. List of fellows. January 1, 1950. In Royal Ent. Soc. London. Proc., Ser. C., J. of Mtgs. 15(1951):sup. to no. 13.
C357 SILBERMANN, G. Énumération des entomologistes vivans, suivie de notes sur les collections entomologiques des principaux musées d'histoire naturelle d'Europe, sur les sociétés d'entomologie, sur les recueils périodiques consacrés a l'étude des insectes, et d'une table alphabétique des résidences des entomologistes. Paris, Roret, 1835. 115 p.
C358 SKINNER, H. The entomologists' directory, containing the names, addresses, special departments of study, etc., of those interested in the study of insect life in the United States and Canada. Philadelphia, 1900. 84 p.
----- Supplement to the Entomologists' directory. Philadelphia, 1904. 25 p.
   One of the earliest directories of North American entomologists.

C359 U.S. BUREAU of ENTOMOLOGY and PLANT QUARANTINE. Directory... 1935-1940. (U.S.D.A. Misc. P. 220 and revisions) 1935-1940.
Earlier directories in this series were issued as U. S. Department of Agriculture Misc. C. no. 80, 1926 and U.S. Department of Agriculture Misc. P. no. 83, 1930.

C360 U.S. BUREAU of ENTOMOLOGY and PLANT QUARANTINE. State leaders, grasshopper and Mormon cricket control. 1948- Washington, 1948-

C361 U.S. BUREAU of ENTOMOLOGY and PLANT QUARANTINE. DIVISION of INSECT IDENTIFICATION. Some of the more important institutions conducting research in entomology; also the more significant entomological societies. Washington, 1946. 33 p.

C362 U.S. EXTENSION SERVICE. List of State extension entomologists and extension apiculturists. September 1949- Washington, 1949-

Handbooks and Texts

C363 U.S. DEPARTMENT of AGRICULTURE. Insects. (Ybk. of Agr., 1952) 780 p.
The most nearly encyclopedic work on economic entomology.
Among its articles on insects and insect control are such reference features as lists of outstanding books, historical material, lists of catalogs of insect groups, and directories of associations and individuals. The bibliographic references appended to many of the articles are selective and suggestive of the most generally used literature on each subject.

C364 BRUES, C. T. Insects and human welfare; an account of the more important relations of insects to the health of man, to agriculture, and to forestry. Rev. ed. Cambridge, Mass., Harvard University Press, 1947. 154 p.

C365 COMSTOCK, J. H. Introduction to entomology. Ed. 9, rev. Ithaca, N. Y., The Author, 1940. 1064 p.

C366 FENTON, F. A. Field crop insects. New York, Macmillan, 1952. 405 p.

C367 FERNALD, H. T. Applied entomology; an introductory textbook of insects in their relations to man. Ed. 5. New York, McGraw-Hill, 1955. 385 p.

C368 GRAHAM, S. A. Forest entomology. Ed. 3. New York, McGraw Hill, 1952. 351 p.

C369 HERMS, W. B. Medical entomology; with special reference to the health and well-being of man and animals. Ed. 4. New York, Macmillan, 1950. 643 p.

C370 METCALF, C. L. and FLINT, W. P. Destructive and useful insects, their habits and control. Ed. 3. New York, McGraw-Hill, 1951. 1071 p.

C371 PAINTER, R. H. Insect resistance in crop plants. New York, Macmillan, 1951. 520 p.

C372 PETERSON, A. A manual of entomological techniques. Ed. 7. Columbus, Ohio, 1953. 367 p.

C373 SCHROEDER, C. W. M. Handbuch der entomologie. Jena, Fischer, 1925-1929. 3 v.
V. 3 contains a chapter on the history of entomology with bibliography, a list of bibliographies useful to the entomologist, a list of entomological journals, etc.
C374 SWAIN, R. B. Insect guide; orders and major families of North American insects. New York, Doubleday, 1948. 261 p.
C375 WILSON, H. F. The historical development of insect classification. St. Louis, Swift, 1937. 133 p.

History

In addition to material in encyclopedias and in general histories of science, biology, zoology, and agriculture, a number of separate histories of entomology are available. Much literature can also be found in journals: local history, reminiscences, historico-bibliographies, and histories of phases of entomology. The subject index of the Bibliography of agriculture (A24), under the heading "Entomology - History," is one guide to this type of publication. Bibliographies and citations in the following books and articles will suggest many additional sources.

C376 BODENHEIMER, F. S. Materialien zur geschichte der entomologie bis Linné. Berlin, Junk, 1928-1929. 2 v.
Extensive bibliographies and surveys of literature.
C377 EISELT, J. N. Geschichte, systematik und literatur der insectenkunde, von den ältesten zeiten bis auf die gegenwart. Leipzig, Hartmann, 1836. 255 p.
Contents: I, Literärgeschichte und systematik der entomologie; II, Literatur der entomologie.
C378 ESSIG, E. O. A history of entomology. New York, Macmillan, 1931. 1029 p.
Emphasizes history of entomology in the Western United States.
C379 HOWARD, L. O. A history of applied entomology (somewhat anecdotal). Washington, Smithsonian Institution, 1930. 564 p.
An informal history of economic entomology and entomologists. About 250 photographs of individuals.
C380 OSBORN, H. A brief history of entomology, including time of Demosthenes and Aristotle to modern times. Columbus, Ohio, Spahr & Glenn, 1952. 303 p.
Anecdotal.
C381 OSBORN, H. Fragments of entomological history, including some personal recollections of men and events. Columbus, Ohio, 1937-1946. 2 v.
Informal and anecdotal, with history of societies and government agencies.
C382 WEBER, G. A. The Bureau of Entomology; its history, activities, and organization. (Inst. for Govt. Res. Serv. Monog. of the U. S. Govt. no. 60.) Washington, Brookings Institute, 1930. 177 p.
C383 WEBER, G. A. The Plant Quarantine and Control Administration; its history, activities, and organization. (Inst. for Govt. Res. Serv. Monog. of the U. S. Govt. no. 59.) Washington, Brookings Institute, 1930. 198 p.

C384 WEISS, H. B. The pioneer century of American entomology. New Brunswick, N. J., 1936. 320 p.
To 1865.

Biography

Biographical sketches of leading entomologists are included in the various editions of standard biographical reference works, such as the American men of science (A83) and Who's who in America. The following Bibliographies of biographical material on entomologists supplement each other.

C385 WADE, J. S. A bibliography of biographies of entomologists, with special reference to North American workers. In Ent. Soc. Amer. Ann. 21(1928):489-520.
C386 CARPENTER, M. M. Bibliography of biographies of entomologists. In Amer. Midland Nat. 33(1945):1-116.
C387 CARPENTER, M. M. Bibliography of biographies of entomologists (Sup.). In Amer. Midland Nat. 50(1953):257-348.

Geography

C388 COMMONWEALTH INSTITUTE of ENTOMOLOGY. Distribution maps of insect pests. Series A. Map no. 1- June 1951- London, 1951-
A currently published geographic guide to insect populations.

Periodical Lists

Chamberlin (C290) includes an annotated list of journals of interest to entomologists. The U.S. Department of Agriculture's Catalogue of publications relating to entomology (C308) includes an extensive list of serials received in the U.S. Department of Agriculture Library in 1906. Lists of journals will be found in many of the older bibliographies: Percheron (C302), Hagen (C299), and Horn (C300). Many of the histories of entomology describe the important serial publications historically. The Entomological news, in its January and June issues, lists the journals it indexes.

C389 CLAASSEN, E. S. An annotated check list of the more important entomological periodicals. In Ent. Soc. Amer. Ann. 38(1945):403-411.
C390 COMMONWEALTH INSTITUTE of ENTOMOLOGY. The review of applied entomology. Serial publications examined. London, 1953. 25 p.
Lists serials examined for abstracts to be published in Review of applied entomology (C219 and C293).
C391 KORSCHEFSKY, R. Verzeichnis der sammelindices von zeitschriften mit mehr oder weniger entomologischen inhalt. In Arb. Morph. Taxonom. Ent. Berlin-Dahlem. 7(1940):209-225.
C392 ROYAL ENTOMOLOGICAL SOCIETY of LONDON. List of serial publications in the Society's library. London, 1951. 28 p.

Miscellaneous

C393 BANKS, N. Directions for collecting and preserving insects. (Smithsonian Inst. U.S. Natl. Mus. B. 67) 1909. 135 p.

C394 BRUES, C. T. Classification of insects; keys to the living and extinct families of insects, and to the living families of other terrestrial arthropods. Ed. 2. (Harvard U. Mus. of Compar. Zool. B. 108.) Cambridge, Mass., 1954. 917 p.

Useful, not only as a toxonomic guide to insects, but as a bibliography of catalogs and descriptive literature on the various families and other groups of insects.

C395 DAVIS, J. J. The entomologist's joke book. Lafayette, Ind., Exterminator's Log, 1937. 160 p.

C396 DRAKE, C. J. Bibliography of the state plant pest laws, quarantines, regulations and administrative rulings of the United States of America. Ames, Ia., Iowa State College Press, 1942. 60 p.

C397 HORN, W. and KAHLE, I. Über entomologische sammlungen, ein beitrag zur geschichte der entomo-museologie. (Ent. Beih. aus Berlin-Dahlem. v. 2-4, 1935-1937.) Berlin-Dahlem, Friedländer, 1935-1937. 536 p.

Provides a list entomological collections with historical information about past owners, data on museums, and entomomuseology. Important institutional and private collections are also described in Howard (C379), Osborn (C381), and other general writers of the field.

C398 JONES, M. P. 4-H club insect manual. (U.S.D.A. Agr. Handb. 65) 1954. 64 p.

APICULTURE

The genus Apis mellifera, the honeybee, is the subject of most of the literature devoted to bees. The apiculturist has a literature of his own with its own guides and reference sources; the entomologist who is a bee specialist shares the literature of general entomology and its reference sources. Many of the publications mentioned in the section on economic entomology include material of interest to professional apiculturists also. Biological abstracts (A11), the Review of applied entomology (C219 and C293), and the Experiment station record (A22) also include bee literature, and relatively complete indexing is provided by the Bibliography of agriculture (A24). The Agricultural index (A23) lists selected publications issued in the United States.

Bibliographies of Bibliographies

C399 FRYKHOLM, L. Översikt över viktigare bilitterature, bibliografiska hjälpmedel och forsknings-institutioner. In Bitidningen 1946:11-14.

Briefly summarizes the scope of bee literature and its bibliographical equipment.

Abstracting Journals

C400 APICULTURAL abstracts. v. 1-   1950-
Published monthly in the Bee World. Also issued separately each month as a reprint. Covers over 300 journals monthly. Annual author and subject indexes.

C401 ARCHIV für bienenkunde. v. 1-   1919-   Lindau, etc., 1919-
Issued irregularly. Includes an abstract section of world bee literature, "Neuere Bienenkunde des In- und Auslandes." The abstracts are long and detailed and arranged by broad subjects.

Bibliographies

C402 BROSCH, A. Verzeichnis der bienenliteratur des in- und aus-landes 1890-1918. In Arch. f. bienenkunde. 1(1919):241-315.

C403 FISCHER, V. E. List of publications on apiculture contained in the U.S. Department of Agriculture Library and in part those contained in the Library of Congress. (U.S.D.A. Libr. Bibliog. Contrib. 21) 1930. 218 p.
Separate lists of books, reprints, and separates; U.S. Department of Agriculture publications; publications of states, territories, and insular possessions; laws; periodicals arranged by title and by country. An author index, but no subject index.

C404 KELLER, A. de. Elenchus librorum de apium cultura. Bibliographie universelle d'apiculture. Milan, Hoepli, 1881. 223 p.

C405 SCOTTISH BEEKEEPERS' ASSOCIATION. Catalogue of the Moir Library. Dundee, 1950. 193 p.

C406 SOCIÉTÉ CENTRALE d'APICULTURE et d'INSECTOLOGIE. Catalogue de la bibliothèque ... Lille, 1896. 78 p.

C407 SOCIÉTÉ d'APICULTURE d'ALSACE-LORRAINE. Katalog der bücher und schriften der central-bibliothek ... Ed. 3. Strassburg, Fischbach, 1909. 27 p.

C408 SOCIÉTÉ ROMANDE d'APICULTURE. BIBLIOTHÈQUE. Catalogue. Fribourg, Switzerland, 1949. 29 p.

C409 THEILER, J. Katalog der bibliothek des Vereins deutschschweizerischer bienenfreunde, mit nachtrag 1941. Zug, Rosenberg, 1937.
Reissue with supplement; imprint gives original publication date.

C410 WALKER, H. J. O. Catalogue of bee books collected and offered for sale by Lt.-Col. H. J. O. Walker. Exeter, Eng., Townsend, 1929. 144 p.

C411 WISCONSIN. UNIVERSITY. COLLEGE of AGRICULTURE. LIBRARY. A list of the publications on apiculture contained in the Dr. Charles C. Miller Memorial Apiculture Library. Madison, Wis., 1936. 283 p.
This includes the celebrated Walker collection, also listed in (C410). Contains a long list of bee journals, with some historical annotations.

Encyclopedic Works

C412 ALPHANDÉRY, E. Encyclopédie apicole. Paris, Baillière, 1945-1948. 4 v.

C413 JACOBY, R. Das imker ABC: nachschlagewerk für alle gebiete der bienenzucht. Segeberg, Die Bienenzucht, 1949. 768 p.
C414 ROOT, A. I. and E. R. The ABC and XYZ of bee culture; an encyclopedia pertaining to scientific and practical culture of bees. Ed. 30. Medina, Ohio, 1954. 703 p.
Approximately 1,000 definitions and explanatory articles.
C415 U.S. BUREAU of ENTOMOLOGY and PLANT QUARANTINE. Information about bee culture. (E-276, rev.) 1953. 10 p.
Includes a directory of apicultural societies, sources of bee information in the United States, and a list of books on beekeeping.

Dictionaries

Since the eighteenth century there has been a steady output of dictionaries of apiculture, most of them more or less encyclopedic. Root's ABC and XYZ of bee culture (C414) defines and gives encyclopedic information concerning apicultural terms.

C416 BELTRAMINI DE CASATI, F. Vocabolario apistico italiano e dizionario d'apicoltura. Milan, Guigoni, 1890. 376 p.
C417 CHRIST, J. L. Algemeines theoretisch-praktisches wörterbuch über die bienen und die bienenzucht. Frankfurt, 1807. 447 p.
C418 CRANE, E. E. Dictionary of bee-keeping terms, with allied scientific terms, giving translations from and into English, French, German, Dutch, with Latin index. London, Bee Research Association, 1951. 74 p.
About 1,000 terms defined.
C419 HANDWÖRTERBUCH für bienenfreunde. Eine alphabetisch geordnete zusammenstellung der ergebenisse der neuesten forschungen über naturgeschichte, pflege und ausnutzung der bienen, so wie des wissenswerthesten aus der geschichte der bienenzucht und ihrer freunde bis zur gegenwart. Berlin, Heinicke, 1867. 264 p.
C420 KIRSTEN, G. Vollständiges worterbuch der bienenkunde und bienenzucht. Ed. 2. Weimar, 1858. 297 p.
C421 KRANCHER, O. Kleines lexikon der bienenzucht und bienenkunde, unter teilweiser berucksichtigung von geschichte und pflanzenkunde für bienenzüchter. Ed. 2. Leipzig, Felt, 1908. 507 p.
C422 OVERBECK, J. A. Glossarium melitturgigum, oder bienenwörterbuch, in welchem die bisher bey der bienenpflege bekannt gewordene oder gebräuchliche kunstwörter und redensarten nach alphabetischer ordnung erklärt werden. Bremen, Forster, 1764. 152 p.
C423 PHIN, J. A dictionary of practical apiculture. New York, Industrial, 1884. 80 p.
C424 PUHLEMANN, K. Imkers wortschatz. (Deutsche Bienenztg. Kleine D.B.Z. bücher. Nr. 8.) Berlin, Deutsche Bienenzeitung, 1948. 111 p.

Directories

C425 ALLEY, H. The national beekeepers' directory. Containing a classified list of the beekeepers of the United States and Canada; with essays and hints regarding the successful management of the apiary. Salem, Mass., Salem Press, 1889. 139 p.

C426 U.S. BUREAU of ENTOMOLOGY and PLANT QUARANTINE. DIVISION of BEE CULTURE. List of dealers in beekeeping supplies, package bees, and queens. (E-297, rev.) Beltsville, Md., 1950. 15 p.
Revised at irregular intervals.

C427 WORLD list of bee research workers. 1953-   London, Bee Research Association, 1953-
Issued irregularly. 1953 ed. lists 370 research workers in 50 countries. Also lists bee research agencies. "First of its kind."

History

C428 FRASER, H. M. Beekeeping in antiquity. Ed. 2. London, University Press, 1951. 145 p.

C429 PELLETT, F. C. History of American beekeeping. Ames, Ia., Iowa State College Press, 1938. 213 p.
Has chapters on the history of journals, books, etc.

C430 RANSOME, H. M. The sacred bee in ancient times and folklore. Boston, Houghton Mifflin, 1937. 308 p.

Periodical Lists

Annotated lists of bee journals are contained in Fischer (C403), with supplementary lists by country. The C. C. Miller Memorial Library list (C411) also includes a separate list of journals with some annotations. Pellett (C429) gives historical background material on American bee journals.

C431 A CATALOG list of American bee journals, including Canada and the United States, with a list of the numbers and volumes in the Miller Memorial Library, July 1, 1930. In Iowa State Hort. Soc. Rpt. 65(1930):462-471. 1930.

C432 FRYKHOLM, L. Förteckning över bitidskrifter. In Bitidningen 49(1950):8-11, 39-45; 50(1951):139-142.
Lists about 380 journals published throughout the world.

C433 MILUM, V. G. Information on beekeeping journals. Urbana, Ill. 1954. 12 p.
Information on dates, irregularities of publications, changes, dates of final issues, etc.

PEST CONTROL

Pest control is concerned not only with insecticides and insect-control techniques but also with rodenticides and control devices for other forms of animal predators on economic crops and livestock. Pest-control literature covers such topics as organic and inorganic insecticides, use of insects or other forms of animal life as control devices, and mechanical equipment for pest control. Chemical literature, which predominates, can best be located by means of guides to chemical literature, which are listed in the section on agricultural chemistry.

Abstracting Journals

Chemical abstracts (D8) provides an extensive coverage of the literature of insecticides and rodenticides, including applications to domestic crops and livestock. Some literature on these topics is also covered by Biological abstracts (A11), Horticultural abstracts (B83), and Field crop abstracts (B84). Articles on mechanical aspects of applications, etc., are abstracted in Agricultural and horticultural engineering abstracts (D55). Both sections A and B of the Review of applied entomology (C219, C293) abstract heavily the literature of insect control by chemicals. The Experiment station record (A22) abstracted and carried references to a great deal of literature on insecticides and their agricultural uses and is particularly useful in locating older literature.

C434 GREAT BRITAIN. COLONIAL OFFICE. COLONIAL PESTICIDE RESEARCH SERVICE. Pesticides abstracts and news summary. Section A, Insecticides, v. 1-   Feb. 1955-   ; Section C, Herbicides, Arboricides, Defoliants, v. 1-   Apr. 1955- London, 1955-
   Section B (apparently not issued by April 1957) is to be concerned with Herbicides. Preceded by: Great Britain. Agricultural Research Council. Fungicide and Insecticide Research Co-ordination Service. Abstracts and news summary. Section A, Insecticides, no. 1-28, 1947-1953; Section C, Herbicides, no. 1-9, 1951-1953.
C435 PEST control abstracts. 1-   1950-   Calcutta, Imperial Chemical Industries (India), Fertilizer and Pest Control Department, 1950-
   Issued annually.

Bibliographies and Indexes

The standard agricultural bibliographic services, Bibliography of agriculture (A24) and the Agricultural index (A23) cover the literature of agricultural applications of insecticides and other insect- and pest-control methods.

C436 GREAT BRITAIN. COLONIAL PRODUCTS LABORATORY. CONSULTATIVE COMMITTEE on INSECTICIDE MATERIALS. Bibliography of insecticide materials of vegetable origin. 35- April/June 1946-
   Quarterly. No. 1-34 appeared in each quarterly issue of the Bulletin of the Imperial Institute, beginning with 36(1938):123-127. Issued separately beginning with no. 35.
C437 HAWES, I. L. Bibliography on aviation and economic entomology. (U.S.D.A. Bibliog. B. 8) 1947. 186 p.
   Revises and brings up-to-date previous bibliographies on the same subject published in 1928 and 1934. Covers literature from 1919-1944.

C. Animal Sciences

C438 McINDOO, N. E. Plants of possible insecticidal value; a review of the literature up to 1941. (U.S. Bur. Ent. and Plant Quarantine E-661) 1945. 286 p.
  The E-series of the Entomology Research Branch, U.S. Agricultural Research Service, contains a great many bibliographies on specialized subjects on insect control and insecticidal investigations. This is an example.

C439 NATIONAL RESEARCH COUNCIL. CHEMISTRY SUBCOMMITTEE. Annotated bibliography of analytical methods for pesticides. Prep. for the Food Protection Committee, Food and Nutrition Board, National Research Council. (Natl. Res. Council. P. 241.) Washington, 1952-1954. 2 v.
  Annotated bibliography of particular value for those working on the effect of pesticides on food products.

C440 STORER, T. I. and MANN, M. P. Bibliography of rodent control. (Natl. Res. Council. Insect Control Comt. Rpt. no. 182.) Washington, 1946. p. 1-324, 1-57.

Dictionaries

Most of the terms and chemical names used in insecticiae work are defined, explained, or illustrated in standard chemical dictionaries, encyclopedias, handbooks, and formularies. Problems in terminology are outlined in the following.

C441 BRITISH STANDARDS INSTITUTION. Recommended common names for pest control products. London, 1952. 14 p.

C442 HALLER, H. L. Common names of insecticides for use in the Journal of Economic Entomology. In J. Econ. Ent. 48(1955):112-115.

Directories

Insecticide research laboratories are listed in the National Research Council's Industrial research laboratories of the United States (A72). The Pesticide handbook (C457) contains a list of commercial pest-control operators and names of manufacturers. The Official publication (C446) of the Association of Economic Poisons Control Officials, Inc. contains a directory of economic-poisons-control officials.

C443 ASSOCIATION of BRITISH INSECTICIDE MANUFACTURERS. Directory. 1948-   London, 1948-
  Issued irregularly (1948, 1950, 1956). 1956 edition has cover title, British insecticides, fungicides and weedkillers.

C444 ENTOMA; a directory of insect and plant pest control, 1935- College Park, Md., Entomological Society of America, 1935-
  1953/54 ed. is 10th. Lists insecticide manufacturers, machinery, fumigation consultants, supplies and suppliers, government officials, pest control operators, trade-marked insecticides, societies, etc.

C445 SOAP and chemical specialities. Blue book and catalog section. New York, McNair-Dorland, 1956. 268 p.
  Annual. Includes insecticides and insect control devices.

Handbooks, Manuals, Texts, etc.

The following are representative of the great number of handbooks and encyclopedic works of various kinds devoted to insecticides.

C446 ASSOCIATION of AMERICAN PESTICIDE CONTROL OFFICIALS.
Pesticide official publication and condensed data on pesticide chemicals. 1953-   College Park, Md., 1953-
Annual. Includes a summarized compilation of chemical, physical, and biological properties of about 150 pesticide chemicals, showing trade names, chemical names, formulas, properties, uses, applications, and storage directions.

C447 BAILEY, S. F. and SMITH, L. M. Handbook of agricultural pest control. New York, Industry Publications, 1951. 191 p.

C448 BROWN, A. W. A. Insect control by chemicals. New York, Wiley, 1951. 817 p.

C449 CANADA. DEPARTMENT of AGRICULTURE. SCIENCE SERVICE.
Guide to the chemicals used in crop protection. London, Ont., 1952. Loose-leaf.

C450 DE ONG, E. R. Insect, fungus, and weed control. New York, Chemical, 1953. 400 p.

C451 DETHIER, V. G. Chemical insect attractants and repellents. New York, Blakiston, 1947. 289 p.

C452 FREAR, D. E. H. A catalogue of insecticides and fungicides.
(Ann. Cryptogamici et Phytopath. v. 7-8.) Waltham, Mass., Chronica Botanica, 1947-1948. 2 v.
Includes bibliographic citations to source articles and patents.

C453 FREAR, D. E. H. Chemistry of the pesticides. Ed. 3. New York, Van Nostrand, 1955. 469 p.
Extensive bibliographies.

C454 INSECTICIDES. In U.S.D.A. Ybk. of Agr., 1952:748-750.

C455 MERCK index of chemicals and drugs. Ed. 6. Rahway, N. J., 1952. 1167 p.
Includes insecticides, giving formulas, characteristics, uses, commercial availability, some literature references, etc.

C456 NATIONAL RESEARCH COUNCIL. CHEMICAL-BIOLOGICAL COORDINATION CENTER. Summary tables of biological tests.
v. 1-   October 1949-   Washington, 1949-
Bimonthly. The Chemical-Biological Coordination Center was established in 1946 to succeed the wartime Insect Control Committee. Its chief purpose is to test and report on the biological effect of certain chemicals. Results are reported in these tables. Particular attention is given to the testing of pesticides.

C457 PESTICIDE handbook. 1951-   State College, Pa., 1951-
Issued annually. Alphabetic list of trade names, list of products according to use, list of commercial pest-control operators, names of manufacturers, etc.

C458 SHEPARD, H. S. The chemistry and action of insecticides. New York, McGraw-Hill, 1951. 504 p.

C459 U. S. AGRICULTURAL RESEARCH SERVICE. ENTOMOLOGY RESEARCH BRANCH. Chemicals evaluated as insecticides and repellents at Orlando, Fla. (U.S.D.A. Agr. Handb. 69) 1954. 397 p.
Evaluates about 11,000 compounds for possible use against insects.

C460 WORLD HEALTH ORGANIZATION. Insecticides, manual of specifications for insecticides and for spraying and dusting apparatus. Geneva, 1953. Loose-leaf.

Legislation

There is a considerable literature concerning the laws and regulations governing the manufacture, sale, and use of insect control methods. Model laws and a summary of state regulations are included in the Official publication of the Association of Economic Poisons Control Officials (C446).

C461 SMITH, C. L. Guide to the laws affecting insecticides, fungicides, and related products. New York, Agricultural Insecticide and Fungicide Association, 1946. Loose-leaf.
   This service provides copies of federal, state, and local laws and regulations, with news and interpretations.

C462 U.S. AGRICULTURAL RESEARCH SERVICE. Regulations for the enforcement of the Federal insecticide, fungicide, and rodenticide act. Regulations of the Secretary of Agriculture; joint regulations of the Secretary of Agriculture and the Secretary of the Treasury for the enforcement of section 10 of the act. Federal insecticide, fungicide, and rodenticide act (61 Stat. 163:7 U.S.C. 135-135k). Domestic regulations effective November 1, 1947; joint regulations with respect to imports effective January 30, 1948. Issued March 1948, reissued with amendments March 1954. (Serv. and Regulat. Announce. 166) 1954. 18 p.
   Previous editions issued by the U.S. Production and Marketing Administration.

C463 WEBER, G. A. Food, drug, and insecticide administration; its history, activities, and organization. (Inst. for Govt. Res. Serv. Monog. of the U.S. Govt. no. 50.) Baltimore, Johns Hopkins Press, 1928. 134 p.
   A history of earlier government activities in administering insect regulation.

COMMERICAL FISHING AND FISHERIES

This section is restricted to literature dealing with the economic exploitation of aquatic animal life. A few obvious references to the literature of fishing as a sport are included, but only as they add to the reference source material of economic fishing. The range of subjects includes such specifically economic aspects as fishery management and the broader subject bases upon which it rests: marine biology, oceanography, limnology, and ichthyology, for example.

Bibliographies of Bibliographies

C464 FOOD and AGRICULTURE ORGANIZATION of the UNITED NATIONS. Fisheries technology literature. (Sup. to World Fisheries Abstracts, v. 3) 1952. 48 p.

Books and other publications in the field of fisheries technology listed with annotations. A bibliography, rather than a guide to reference books, it nevertheless cites literature that is useful in locating information. Printed so that references can be cut out and filed in standard card trays.

C465 SCATTERGOOD, L. W. Bibliographic sources for fishery students and biologists. In Amer. Fisheries Soc. Trans. 83(1953):20-37.
Annotated list of bibliographic journals, abstracting journals, journals which include bibliographic and abstract sections, important general works on hydrobiology, fisheries and general zoology, special bibliographies, general bibliographies, and indexes to reports on fisheries and allied subjects.

C466 KEEFE, R. V. Sources of information concerning the commercial fisheries. (U. S. Fish and Wildlife Serv. Fishery L. 362) 1952. 23 p.
Though chiefly a guide to the technical literature on various phases of commercial fisheries, such as canning, consumption, cookery, salting, smoking, spoilage, etc., also lists sources of directory information, technical journals, trade journals, and visual aids.

Abstracting Journals

General zoological abstracting journals, particularly Biological abstracts (A11), cover the literature of marine biology, systematic ichthyology, and limnology. (See Section G, Addendum, for additional material.)

C467 U.S. FISH and WILDLIFE SERVICE. Commercial fisheries abstracts. 1-   1947-   Washington, 1947-
Monthly "summary of important developments for the fishery industries." Abstracts are printed in card size to be cut out and filed by subject. In addition to abstracts, there are bibliographical citations to literature of less general interest in each issue. Author and subject indexes in each issue. Triennial indexes, author and subject, have been published for the periods 1948-1950, 1951-1953.

C468 WORLD fisheries abstracts. 1-   1950-   Rome, Food and Agriculture Organization of the United Nations, 1950-
Bimonthly. Mostly republished from Commercial fisheries abstracts (C467). Like the latter, covers journals, laboratory reports, government documents on fishery technology, processing methods, boat design, fishing methods, chemical examination of fishery products, etc. Covers world literature. The following handbook (C469) describes its use and coverage.

C469 FOOD and AGRICULTURE ORGANIZATION of the UNITED NATIONS. Handbook for World Fisheries Abstracts, a bimonthly review of technical literature in fisheries and related industries. Washington, 1950. 155 p.

C470 ZEITSCHRIFT für fischerei und deren hilfswissenschaften. 1-   1893-   Berlin, 1893-
About 6 issues a year. Contains literature reviews.

C. Animal Sciences

Bibliographies and Indexes - General

The most comprehensive indexing service of the literature of ichthyology, especially its biological and taxonomic aspects, is provided by the Zoological record (C10). For extensive lists of bibliographies see also (C465), and (C464). (See Section G, Addendum, for additional material.)

C471 BIBLIOGRAPHIA oceanographica. 1- 1928- Venice, Colegio Talassographici Italici, 1929-
Covers world literature on physical oceanography, biological oceanography, fishing, etc. Under fishing, there are sections on fisheries economy and legislation, application of biology to fishing, fishing ports, fishing methods and gear, fishing localities, products of the fisheries, nutritional value of fishery products, markets, fishery products industries and commerce, and preservation.

C472 DEAN, B. A bibliography of fishes. New York, American Museum of Natural History, 1916-1923. 3 v.

C473 INTERNATIONAL COUNCIL for the EXPLORATION of the SEA. Journal du conseil. 1- 1926- Copenhagen, 1926-
Issued three times a year. "Current bibliography" in each issue lists world literature on oceanography, hydrography, plankton, organic production, bottom fauna, fish, marine fisheries, freshwater fisheries, shell fisheries, etc.

C474 LICHTENFELT, H. F. W. Literatur zur fischkunde. Bonn, Hager, 1906. 140 p.

C475 PARIS. STATION CENTRALE d'HYDROBIOLOGIE APPLIQUÉE. Analyses bibliographiques, 1940-1950. (Special unnumbered issue of its Annales.) Paris, 1953. 175 p.
A bibliography of French publications in hydrobiology for the period 1940-1950.

Bibliographies and Indexes - Early

C476 MULDER BOSGOED, D. Bibliotheca ichthyologica et piscatoria. Catalogus van boeken en geschriften over de natuurlijke geschiedenis van de visschen en walvisschen, de kunstmatige vischteelt, de visscherijen, de wetgeving op de visscherijen. Haarlem, Loosjes, 1874. 474 p.

Bibliographies and Indexes - Special

C477 APPLEGATE, V. C., MACY, P. T., and HARRIS, V. E. Selected bibliography on the applications of electricity in fishery science. (U.S. Fish and Wildlife Serv. Spec. Sci. Rpt.: Fisheries 127) 1954. 55 p.

C478 INGRAM, W. M. Publications on industrial wastes relating to fish and oysters; a selected bibliography. (U.S. Pub. Health Serv. Div. of Water Pollut. Control. Pub. Health Serv. P. 270) 1953. 28 p.

C479 INTERNATIONAL COUNCIL for the STUDY of the SEA. Index to publications by the International Council for the Exploration of the Sea, 1899-1938. Copenhagen, Høst, 1939. Loose-leaf.

Commercial Fishing 121

C480 JENKINS, J. T. Bibliography of whaling. In Soc. Bibliog. Nat. Hist. J. 2(1948):71-166.
C481 MACIOLEK, J. A. Artificial fertilization of lakes and ponds; a review of the literature. (U.S. Fish and Wildlife Serv. Spec. Sci. Rpt.: Fisheries 113) 1954. 41 p.
C482 MORTIMER, C. F. Fertilizers in fishponds. (Gt. Brit. Colonial Off. Fishery P. 5.) London, 1954. 155 p.
    A review of the literature from 1900-1941 on the fertilization of ponds to provide food sources for fish.
C483 U.S. FISH and WILDLIFE SERVICE. Doctoral dissertations on the management and ecology of fisheries. (Spec. Sci. Rpt.: Fisheries 31) 1950. 35 p.
----- Supplement. (Spec. Sci. Rpt.: Fisheries 87) 1952. 44 p.
    Lists dissertations accepted between 1934 and 1951.

Bibliographies and Indexes - United States Government Publications

A summary of the history of various federal government agencies concerned with fisheries is given in A. M. Boyd's United States government publications (N. Y., Wilson, 1949). Various bibliographies of publications of these agencies have been compiled.

C484 U.S. BUREAU of FISHERIES. An analytical subject bibliography of the publications of the Bureau of Fisheries, 1871-1920, by Rose M. E. MacDonald. (Doc. 899) 1921. 306 p. (Also as its Rpt. of the Cmnr., 1919/20. App. V.)
    See (G19) and (G20) for later indexes.
C485 U.S. FISH and WILDLIFE SERVICE. Available leaflets on fisheries, 1956. (Fisheries L. 9) 1956. 17 p.
    Revised at frequent intervals.
C486 U.S. FISH and WILDLIFE SERVICE. Monthly list of printed and duplicated material. Washington, 1955-
C487 U.S. SUPERINTENDENT of DOCUMENTS. Fish and wildlife. Ed. 43. (Price List 21) 1955. 10 p.
    Revised approximately annually. Lists only in-print publications available for sale.

Bibliographies and Indexes - Angling

C488 FEARING, D. B. Check list of books on angling, fish, fisheries, fish-culture, etc. New York, 1901. 138 p.
C489 HAMPTON, J. F. Modern angling bibliography; books published on angling, fisheries, fish culture, from 1881-1945. London, Jenkins, 1947. 99 p.
C490 NISSEN, C. Schöne fischbücher. Stuttgart, Hempe, 1951. 108 p.
    Forty-six additional titles to this list are proposed in an article entitled "Illustrated works on fishes," by John C. Briggs, that appeared in Copeia 3(1955):243-245.
C491 NEW YORK PUBLIC LIBRARY. List of works in the New York Public Library relating to fishing and fish culture. New York, 1909. 49 p.
    (Reprinted from the N. Y. Pub. Libr. B. 13. 1909.)

C492 ROBB, J. Notable angling literature. London, Jenkins, 1946. 229 p.
C493 WESTWOOD, T. Bibliotheca piscatoria; a catalogue of books on angling, the fisheries, and fish-culture, with bibliographical notes and an appendix of citations touching on angling and fishing from old English authors. London, Satchell, 1883. 397 p.
C494 WESTWOOD, T. A list of books relating to fish, fishing, and fisheries to supplement the Bibliotheca piscatoria ... This list includes works published between 1883 and 1900, and some older books not mentioned in the "Bibliotheca piscatoria." In Eng. Cat., v. 6, 1898-1900. London, Low, 1901. App. C, p. 751-776.
C495 WETZEL, C. M. American fishing books; a bibliography from the earliest times up to 1948, together with a history of angling and angling literature in America. Newark, Del., 1950. 235 p.

Bibliographies and Indexes - Translations

C496 SCATTERGOOD, L. W. English translations of fishery literature. (U.S. Fish and Wildlife Serv. Spec. Sci. Rpt.: Fisheries 35) 1951. 65 p.
----- Supplement (Spec. Sci. Rpt.: Fisheries 176) 1956. 32 p.
C497 WASHINGTON (STATE). DEPARTMENT of FISHERIES. Translations of fisheries literature from foreign languages into English; comp. by Wilbert McLeod Chapman. Seattle, 1939-1940. 2 v.

Bibliographies and Indexes - Motion Pictures

C498 FOOD and AGRICULTURE ORGANIZATION of the UNITED NATIONS. SECRETARIAT. FISHERIES DIVISION. Fisheries and related subjects: list of films. Rome, 1953. 59 p.
C499 U.S. FISH and WILDLIFE SERVICE. Fishery motion pictures. (Fishery L. 255) 1954. 4 p.

Dictionaries and Encyclopedias

There are a number of specialized glossaries and other types of nomenclatural guides that can assist in determining word usage and meanings.

C500 ALLYN, R. A dictionary of fishes. Ed. 2. St. Petersburg, Fla., Great Outdoors Association, 1951. 108 p.
C501 AMERICAN FISHERIES SOCIETY. COMMITTEE on COMMON and SCIENTIFIC NAMES of FISHES. A list of common and scientific names of the better known fishes of the United States and Canada. Report presented at the seventy-seventh annual meeting, Denver, Colorado, September 10-12, 1947. (Spec. Rpt. no. 1.) Ann Arbor, Mich., 1948. 45 p.
C502 BIBLIOGRAPHIA oceanographica. Vocabulorum elenchus prior lexico totius scientiae oceanographicae redigendo. Venice, 1935. 19 p.
Dictionary of oceanographic terms, with equivalents in Latin, Italian, French, English, and German in parallel columns.

C503 BLY, F. Verklarende vakwoordenlijst van de zee-visscherij. (K. Vlaamsche Acad. v. taal- en letterkunde. Reeks VI, no. 51.) Leuven, De Vlaamsche Drukkerij, 1931. 298 p.
C504 CROKER, R. S. Glossary of Japanese fisheries terms. (Supreme Commander for the Allied Powers. Nat. Resources Sect. Rpt. no. 63.) Tokyo, 1946. 43 p.
C505 FISHERMAN'S encyclopedia. Harrisburg, Pa., Stackpole, 1950. 698 p. Includes bibliographies.
C506 HUET, M. Petit glossaire limnologique. (Centre Belge d'Étude et de Doc. des Eaux. B. nos. 3-4) 1949. 39 p.
Lists limnological and fishery terms in French, German, and English equivalents.
C507 INTERNATIONAL HYDROGRAPHIC BUREAU. Hydrographic dictionary. Ed. 2. (Spec. P. no. 32.) Monaco, 1951. 89 p.
English and French, with equivalents in Danish, Dutch, Portuguese, Spanish, Italian, Norwegian, Swedish, German, etc.
C508 JORDAN, D. S. The genera of fishes ... A contribution to the stability of scientific nomenclature. (Leland Stanford Junior U.P., U.Ser. 27, 36, 39, and 43.) Stanford University, Calif., 1917-1920. 4 v.
C509 PARDO GARCÍA, L. Diccionario de ictiología, piscicultura y pesca fluvial, incluyendo la legislación y la organización e historia del Servicio Piscícola. (His De Rebus Hidrobiologicis, 17.) Madrid, Instituto Forestal de Investigación y Experiencia, 1945. 338 p.
Encyclopedic dictionary, in Spanish, of fish culture, fisheries, fresh water fishing, etc.
C510 ROSA, H. Scientific and common names applied to tunas, mackerels, and spear fishes of the world, with notes on their geographic distribution; a progress report on the compilation of scientific and common names of important food fishes. Washington, Food and Agriculture Organization of the United Nations, 1950. 235 p. Bibliography, p. 168-191.
C511 STRÖMBERG, R. Studien zur etymologie und bildung der griechischen fischnamen. Göteborg, Elander, 1943.
C512 THOMPSON, D. W. Glossary of Greek fishes. London, Oxford, 1947. 302 p.
C513 WILLIAMS, A. C. Dictionary of trout flies and of flies for sea-trout and grayling. New York, Macmillan, 1950. 375 p.

Directories and Yearbooks

Membership lists of such organizations as the American Fisheries Society, the Society of Ichthyologists and Herpetologists, International Association of Game, Fish and Conservation Commissioners, the North American Wildlife Conference, American Society of Limnology and Oceanography, etc., provide directory service to the research field.

In its Fishery Leaflets series, the U.S. Fish and Wildlife Service has issued a number of very useful directories: List of fishery cooperatives, no. 292; List of unions, no. 293; List of manufacturers, no. 195; List of fishery associations, no. 254; List of organizations, no. 313; List of canning companies, no. 359; etc. Only a few of the commercial fishery directories and buyers' guides published throughout the world are listed below. (See note on directories, p. 15.)

C514 AMERICAN FISHERIES SOCIETY. List of members, comp. June, 1954. In Trans. 83(1953):435-466.
   Also includes a list of libraries, clubs, dealers, and state fish and conservation agencies which are members of the American Fisheries Society.

C515 AMERICAN INSTITUTE of BIOLOGICAL SCIENCES. ADVISORY COMMITTEE on HYDROBIOLOGY. Directory of hydrobiological laboratories and personnel in North America. Honolulu, University of Hawaii Press, 1954. 324 p.

C516 CANADIAN fisheries annual, 1955. Gardenvale, Que., National Business Publications, 1955. 132 p.
   Has a statistical section, a directory of companies, and a buyers' guide.

C517 FISHERIES year-book and directory: 1955. International reference book and directory of the fishing and fish processing industries. London, British Continental Trade Press, 1955. 444 p.
   Formerly World fisheries year-book and directory. Description of fish varieties, dictionary of common fish names in seven languages, world directory of firms, organizations, trade journals, etc.

C518 FISHING gazette annual review ... 1923-   New York, Fishing Gazette, 1923-

C519 FISHING tackle digest. 1st-   Annual ed. Chicago, Paul, 1946-

C520 FOOD and AGRICULTURE ORGANIZATION of the UNITED NATIONS. SECRETARIAT. FISHERIES DIVISION. Fishery research and educational institutions in North and South America. Washington, 1950. 85 p.
   Includes government organizations, educational institutions, with description of activities, publications, etc.

C521 SOUTH African fishing industry handbook and buyers' guide, 1954. Cape Town, South African Shipping News and Fishing Industry Review, 1954. 244 p.
   Issued annually.

C522 INTERNATIONAL GAME FISH ASSOCIATION. Year book. 1943- New York, 1943-

C523 TURNER, D. B. Professional opportunities in the wildlife field. Washington, Wildlife Management Institute, 1948. 208 p.
   Directory of schools teaching fishery, analysis of courses, etc.

Handbooks, Texts, etc.

C524 CARLANDER, K. D. Handbook of freshwater fishery biology, with the first supplement. Dubuque, Ia., Brown, 1953. 429 p.

C525 DAVIS, H. S. Culture and diseases of game fishes. Berkeley, University of California Press, 1953. 332 p.

C526 HARDY, A. C. Seafood ships. London, Lockwood, 1947. 248 p.
   Descriptive information on commerical fishing boats.

C527 LÜBBERT, H. Die grossen seefischereien der erde, ihre fanggründe und ihre fangprodukte. Hamburg, Girardet, 1950. 104 p.

C528 NATIONAL GEOGRAPHIC SOCIETY. Book of fishes. 1952 ed. Rev. and enl., presenting the better-known food and game fishes and the aquatic life of the coastal and inland waters of the United States. Washington, 1952. 339 p.

Commercial Fishing 125

C529 PETERSON, C. E. Preliminary review of the fishereies of the United States, 1952. (U.S. Fish and Wildlife Serv. Fishery L. 393) 1953. 20 p.
C530 ROUNSEFELL, G. A. Fishery science, its methods and applications. New York, Wiley, 1953. 444 p.
List of fishery journals, p. 405-08; glossary of terms, p. 409-13.
C531 TRESSLER, D. K. Marine products of commerce; their acquisition, handling, biological aspects, and the science and technology of their preparation and preservation. Ed. 2. New York, Reinhold, 1951. 782 p.
C532 U.S. FISH and WILDLIFE SERVICE. Fishery resources of the United States. Washington, Public Affairs Press, 1947. 134 p.
C533 WELCH, P. S. Limnology. Ed. 2. New York, McGraw-Hill, 1952. 538 p.

History

C534 ANELL, B. Contribution to the history of fishing in the southern seas. (Studia Ethnographica Upsaliensia, 9.) New York, Heinman, 1955. 268 p.
C535 BOWER, W. T. History of the American Fisheries Society. In Amer. Fisheries Soc. Trans., 1910, p. 323-358.
C536 GOODE, G. B. The fisheries and fishery industries of the United States. Washington, 1884-1887. 7 v.
See especially Sect. V, History and methods of the fisheries.
C537 GOODSPEED, C. E. Angling in America; its early history and literature. Boston, Houghton Mifflin, 1939. 380 p.
"Checklist of American angling publications related to fresh-water angling in the United States 1660-1900 chronologically arranged," p. 345-65.
C538 HOHMAN, E. P. The American whaleman; a study of life and labor in the whaling industry. New York, Longmans, 1928. 355 p.
C539 INNIS, H. A. The cod fisheries; the history of an international economy. New Haven, Conn., Yale University Press, 1940. 520 p.
C540 JENKINS, J. T. A history of the whale fisheries. London, Whitherby, 1921. 336 p.
C541 JENKINS, J. T. The sea fisheries. London, Constable, 1920. 299 p.
C542 JUDAH, C. B. The North American fisheries and British policy to 1713. (U. of Ill. Studies in the Social Sci. v. 18, no. 3-4.) Urbana, Ill., University of Illinois Press, 1933. 183 p.
C543 MCFARLAND, R. A history of the New England fisheries, with maps. New York, Appleton, 1911. 457 p.
C544 MOORE, S. A. and H. S. History and law of fisheries. London, Stevens, 1903. 446 p.
C545 NORMAN, J. R. A history of fishes. London, Benn, 1931. 463 p.
C546 RADCLIFFE, W. Fishing from the earliest times. Ed. 2. New York, Dutton, 1926. 494 p.
C547 THOMAZI, A. A. Histoire de la pêche des âges de la pierre a nos jours. Paris, Payot, 1947. 644 p.
C548 TOWER, W. S. A history of the American whale fishery. Philadelphia, University of Pennsylvania Press, 1907. 145 p.
C549 U.S. BUREAU of the CENSUS. Fisheries of the United States. 1908. Washington, 1911. 324 p.

C. Animal Sciences

This and similar Census Bureau publications are chiefly statistical, but may have historical material.

C550 U.S. BUREAU of FISHERIES. The United States Bureau of Fisheries, its establishment, functions, organization, resources, operations, and achievements. Washington, 1908. 80 p.
"Publications," p. 53-56.

Geography

The geography of aquatic life is included in Bartholomew (C51).

C551 EKMAN, S. P. Zoogeography of the sea. London, Sidgwick, 1953. 417 p.
A basic text in the field of marine zoogeography, with a very extensive bibliography.

Periodical Lists

Following are two of the lists of journals in the field of fisheries. Additional lists will be found in Rounsefell (C530), Keefe (C466), in the "List of publications regularly abstracted" in Commercial fisheries abstracts (467), in the first issue of each volume of World fisheries abstracts (C468), and in Fisheries year-book and directory (C517).

C552 BIBLIOGRAPHIA oceanographica. Publications périodiques consultées pour la préparation de la bibliographie. Venice, Ferrari, 1930. 24 p.
C553 U.S. FISH and WILDLIFE SERVICE. Partial list of journals and newspapers concerning the fisheries. (Fishery L. 160) 1953. 9 p.
Revised irregularly.

Statistics

Information about statistical source material is also given in Keefe (C466).

C554 PETERSON, C. E. Fishery statistics. (U.S. Fish and Wildlife Serv. Fishery L. 197) 1952. 12 p.
C555 U.S. FISH and WILDLIFE SERVICE. Statistical digest. 1939- Washington, 1942-
Issued irregularly. Formerly called Fishery statistics of the United States.
C556 YEARBOOK of fisheries statistics. Annuaire statistique des pêches. 1947- Washington, Food and Agriculture Organization of the United Nations, 1947-

Miscellaneous

C557 FOOD and AGRICULTURE ORGANIZATION of the UNITED NATIONS. SECRETARIAT. FISHERIES DIVISION. Commodity standards for fisheries products; a digest of current laws and regulations. (F.A.O. Fisheries Study no. 2.) Rome, 1953. 157 p.
  Extensive bibliography, p. 150-157.
C558 LEONARD, L. L. International regulation of fisheries. (Carnegie Endowment for Internatl. Peace. Div. of Internatl. Law. Monog. Ser. no. 7.) Washington, 1944. 201 p.

SECTION D
PHYSICAL SCIENCES

AGRICULTURAL CHEMISTRY

The chemist working with agricultural products must delve into the whole range of chemical literature. There are, however, a few tools such as the A.O.A.C.'s <u>Official methods of analysis</u> (D18) which are intended specifically for such workers: these, and certain general compilations, such as <u>Chemical abstracts</u> (D8), which are essential for any literature search are listed here. The chemist will also be interested in the "Agriculture - General," "Soils and fertilizers," and the "Food and nutrition" sections of this compilation.

For a more detailed review of chemical reference works the following compilations are suggested.

D1   AMERICAN CHEMICAL SOCIETY. DIVISION of CHEMICAL LITERATURE. Literature resources for chemical process industries. (Advn. in Chem. no. 10.) Washington, 1954. 582 p.
   See E2 for annotation.

D2   AMERICAN CHEMICAL SOCIETY. DIVISION of CHEMICAL LITERATURE. Searching the chemical literature. (Advn. in Chem. no. 4.) Washington, 1951. 184 p.

D3   CRANE, E. J., PATTERSON, A. M., and MARR, E. B. Guide to the literature of chemistry. Ed. 2. New York, Wiley, 1957. 397 p.

D4   MELLON, M. G. Chemical publications, their nature and use. Ed. 2. New York, McGraw-Hill, 1940. 284 p.

D5   SOULE, B. A. Library guide for the chemist. New York, McGraw-Hill, 1938. 302 p.

Abstracting Journals and Reviews

D6   ANNUAL review of biochemistry. v. 1-   1932-   Stanford, Calif., Annual Reviews, 1932-
   Of reference value because of the extensive bibliographies appended to the review articles. Has author and subject indexes.

D7   BIEDERMANN'S zentralblatt für agrikulturchemie und rationellen landwirtschaftsbetrieb. v. 1-60, 1872-1931; Abt. A., v. 1-6

(v. 61-66) 1931-1936; Abt. B, v. 1-16, 1929-1944. Leipzig, Akademische Verlagsgesellschaft, 1872-1944.

----- General register. v. 1-55, 1872-1926. Leipsiz, 1901-1928. 3 v.
Title varies. Abt. A, Allgemeiner und referierender teil; Abt. B, Tierernährung. An international abstracting service of value for the period covered. The term agricultural chemistry was interpreted to include soil science, fertilizers, microbiology, and plant physiology, particularly when the emphasis was on the chemistry of an agricultural product. Includes book reviews. Author and subject indexes were provided for each volume.

D8   CHEMICAL abstracts. v. 1-   1907-   Columbus, Ohio, American Chemical Society, 1907 -
Issued semimonthly. The most comprehensive of the abstracting services, indexing approximately 5,236 periodicals from all over the world. Has annual author, subject, and formula indexes. Decennial indexes are available, covering 1907 to 1946. The following sections are widely used by plant scientists and agriculturists: Organic chemistry, Biological chemistry, Foods, Soils and fertilizers, Pesticides and crop-control agents, The fermentation industries.

D9   CHEMISCHES zentralblatt. Vollständiges repertorium für alle zweige der reinen und angewandten chemie ... jahrg. 1-   1830- Berlin, Verlag Chemie, 1830-
Issued weekly. Title and imprint varies. An abstracting journal, systematically arranged, international in scope, covering the whole field of chemistry. Contains sections on agricultural chemistry, plant and animal chemistry, and physiology. Annual author, subject, patent, and formula indexes. Various cumulative indexes covering several years have also been issued.

D10 JAHRESBERICHT für agrikultur-chemie, ed. by F. Mach. v. 1-76, 1858-1933. Berlin, Parey, 1860-1936. 76 v.

----- General register. v. 1-40, 1858-1897. Berlin, Parey, 1879-1899. 2 v.
Title varies. V. 21-76 are Neue Folge 1-Folge 4. Abstracts and indexes books and periodical articles in many languages. Titles are cited in German only. Material is classified and presented under four headings: Plant production, Animal production, Agricultural industries and fermentation, and Research methods. Author and subject indexes.

D11 ROSE, A. Distillation literature, index and abstracts, 1941-   by Arthur and Elizabeth Rose, with the cooperation of the American Chemical Society . . . State College, Pa., Applied Science Laboratories, c1948-
Issued irregularly. Abstracts of world literature relating to distillation. Of interest to agricultural chemists, food technologists, and other workers with chemical processes. In 1955 won the American Library Association's Oberly Memorial Fund award for the best bibliography on agriculture or related subjects published during 1953-1954.

Bibliographies and Indexes

D12 BEILSTEIN, F. Handbuch der organischen chemie. 4. Aufl. Die literatur bis 1. Januar 1910 umfassend, hrsg. von der Deutschen

Chemischen Gesellschaft, bearb. von Bernhard Prager und Paul Jacobson, unter ständiger Mitwirkung von Paul Schmidt und Dora Stern ... v. 1-   Berlin, Springer, 1918-

----- Ergänzungswerk, die Literatur von 1910-1919 umfassend ... v. 1- Berlin, Springer, 1928-

----- Ergänzungswerk, die Literatur von 1920-1929 umfassend ... v. 1- Berlin, Springer, 1941-

This is the most comprehensive and indispensable index to the literature of organic chemistry. It provides a complete summary of published data on organic compounds. A helpful guide to the use of this work is in Soule's Library guide for the chemist (D5), p. 127-153.

D13   NIKLAS, H., ed. Literatursammlung aus dem gesamtgebiet der agrikulturchemie. München, Verlag des Agrikulturchemischen Instituts Weihenstephan der Technischen Hochschule München, 1931-1939. 5 v.

See D33 for annotation.

D14   U.S. BUREAU of CHEMISTRY and SOILS. Index of publications of the Bureau of chemistry and soils, originally the Bureau of chemistry and the Bureau of soils. 75 years - 1862-1937. Volume 1, list of titles and authors, prep. by H. P. Holman, V. A. Pease, K. Smith, M. T. Reid, and A. Crebassa. Washington, 1939. 546 p.

Arranged chronologically under the following headings: Chemistry publications, Soil publications, Fixed nitrogen research laboratory publications, Food inspection decisions, Public service patents, Soil surveys, Chemistry and soils publications. Includes both official and non-official publications. Author index. Only one volume issued.

Dictionaries and Encyclopedias

D15   ELSEVIER'S encyclopaedia of organic chemistry, ed. by E. Josephy and F. Radt. v. 1-   New York, Elsevier, 1940-

In progress. Comparable to Beilstein (D12). Attempts to cover the literature of organic chemistry up to 1936. Subject and formula indexes in each volume. Twenty volumes are projected, including general subject and formula indexes. Includes some material not in fourth edition of Beilstein and its supplements.

D16   HACKH, I. W. D. Hack's Chemical dictionary (American and British usage). Containing the words generally used in chemistry, and many of the terms used in the related sciences of physics, astrophysics, mineralogy, pharmacy, agriculture, biology, medicine, engineering, etc.; based on recent chemical literature ... Ed. 3, completely rev. and ed. by Grant ... Philadelphia, Blakiston, 1944. 925 p.

Defines more than 57,000 terms and contains numerous tables, diagrams, portraits, and other illustrations.

D17   THORPE, J. F. and WHITELEY, M. A. Thorpe's dictionary of applied chemistry. Ed. 4. New York, Longmans, 1937-1956. 12 v.

Alphabetically arranged dictionary containing many short articles, and occasional long encyclopedia-type articles with bibliographies. Describes and gives information regarding chemical composition of agricultural products. Illustrated. V. 12 is General Index.

Handbooks, Manuals, etc.

D18 ASSOCIATION of OFFICIAL AGRICULTURAL CHEMISTS. Official methods of analysis. Committee on editing methods of analysis, H. J. Fisher [and others] Ed. 7, 1950. Washington, 1950. 910 p.

Methods of analysis of soils, fertilizers, insecticides, agricultural products, etc. Standard solutions, reference tables and definitions of terms, fertilizers, and liming materials are included. Indexed.

D19 BENNETT, H., ed. The chemical formulary; a collection of valuable, timely, practical commercial formulae and recipes for making thousands of products in many fields of industry. v. 1-9. Brooklyn, Chemical, 1933-1951. 9 v.

Chapters on flavors and beverages, cosmetic and drug products, farm and garden specialities, food products, and textiles and fibers are included. V. 7 contains a directory of sources of supply of trade-name and other chemicals. A cumulative index for v. 1-6 was issued in 1944.

D20 BROWNE, C. A. A source book of agricultural chemistry. (Chron. Bot., v. 8, no. 1) Waltham, Mass., Chronica Botanica, 1944. 290 p.

An historical account of the origins of agricultural chemistry. Gives biographical information and extracts from the writing of early agricultural chemists. Indexed by name and subject. An appendix lists references on the history of agricultural chemistry.

D21 FREAR, D. E. H., ed. Agricultural chemistry, a reference text. New York, Van Nostrand, 1950. 2 v.

A comprehensive text by various experts. Some chapters have useful bibliographies.

SOILS AND FERTILIZERS

Soil science is unusually well documented and indexed. The publications of the Commonwealth Bureau of Soil Science, for instance, cover the subject thoroughly from 1931 to date; the bibliography edited by H. Niklas (D33) and his co-workers is an excellent guide to papers published before 1931.

The more general reference works are also useful to the soil scientist. The Soils and fertilizers section of the Bibliography of agriculture (A24), for example, is useful in checking for recent titles; the Agricultural index (A23) is a quick subject guide to material in English; the Engineering index (D54) and the Industrial arts index (D60) refer to many papers on the engineering aspects of soil mechanics which are not listed in other sources; Chemical abstracts (D8) has a section on soils and fertilizers; Biological abstracts (A11) devotes a portion of each issue to soil bacteriology.

Soil science, a widely distributed periodical, has a useful book review section in each issue; Zeitschrift für Pflanzenernährung, Düngung, und Bodenkunde (D25) has a "Referate" section in each number which abstracts important additions to the literature of soils, plant nutrition, and fertilizers.

Those interested in soil science should pay particular attention to the preceding and closely related section on agricultural chemistry.

Abstracting Journals

D22   COMMONWEALTH BUREAU of SOIL SCIENCE. Soils and fertilizers. v. 1-   1932-   Harpenden, Eng., 1938-
Issued six times a year. Formerly issued by Imperial Bureau of Soil Science. International current bibliography, classified and annotated. Annual subject and author indexes. Bibliography of soil science, fertilizers, and general agronomy (D28), issued by the Imperial Bureau of Soil Science, serves as a cumulative index to this publication.

D23   KOLLOID zeitschrift. v. 1-   July 1906-   Darmstadt, etc., Verlag von Dr. Dietrich Steinkopff, 1906-
Issued monthly. The "Referate" sections cover world literature and include references to colloid studies of interest to the soil scientist and the agricultural chemist. There are annual author and subject indexes.

D24   U. S. AGRICULTURAL RESEARCH SERVICE. SOIL and WATER CONSERVATION RESEARCH BRANCH. Abstracts of recent published material on soil and water conservation. March 1949- Washington, 1949-
Issued irregularly (eight numbers from 1949 to 1953). Designed for those doing soil conservation work. Each issue has a subject and author index. Formerly issued by the U. S. Soil Conservation Service. Preceded by Research summaries, June 1945-February 1948.

D25   ZEITSCHRIFT für pflanzenernährung, düngung und bodenkunde. v. 1- 1922-   Berlin, Verlag Chemie, 1922-
Usually twelve issues a year. A general review of soil science, plant nutrition, and fertilization, which includes a useful book review and abstract section in each issue. A "Generalregister der Originalarbeiten," v. 1-55, n.s., 1936-1951 was issued in 1953. Issued in two parts up to v. 36, 1934 (A, "Wissenschaftlicher teil"; B, "Wirtschaftlich-praktischer teil"). Title was Bodenkunde und Pflanzenernährung from 1936-1945 (n.s., v. 1-36).

Bibliographies and Indexes

D26   CARNEGIE LIBRARY of PITTSBURGH. Bibliography of physical properties and bearing value of soils. In American Society of Civil Engineers Proceedings, 43(1917):1192-1240.
-----Supplement, comp. by Morris Schrero. In Amer. Soc. of Civil Eng. Proc. 57(1931):871-921.
References are to items in Carnegie Library of Pittsburgh.

D27   CHILEAN NITRATE EDUCATIONAL BUREAU, INC. Bibliography of the literature on the minor elements and their relation to plant and animal nutrition. Ed. 4. New York, 1948-1955. 4 v.
Abstracts and references originally were obtained from Chemical abstracts (D8), Experiment station record (A22), and Soils and fertilizers (D22). Has author, element, general nutrition, and botanical indexes.

D28   COMMONWEALTH BUREAU of SOIL SCIENCE. Bibliography of soil science, fertilizers, and general agronomy, 1931-1950. Harpenden,

Eng., 1935-1951. 6 v.

Serves as a cumulative index to Soils and fertilizers (D22), issued by the Commonwealth Bureau of Soil Science. Closely classified title bibliography of world literature. Indexed by author and subject.

D29 FELDKAMP, C. L. and PENNINGTON, C. E., comps. List of the publications on soils issued by the state agricultural experiment stations of the United States through 1926. (U.S.D.A. Libr. Bibliog. Contrib. 15) 1927. 81 p.

Arranged alphabetically by states. Not annotated or indexed.

D30 GAINES, S. H., comp. Bibliography on soil erosion and soil and water conservation ... With abstracts by F. Vincent, M. Bloom, and J. F. Carter. (U.S.D.A. Misc. P. 312) 1938. 651 p.

Classified list of 4,388 references, the majority of which were published during the period 1900-August 1927. Includes list of citations by geographic regions and an author index.

D31 HAWKS, E. B. and TROLINGER, C., comps. List of the publications on soils issued by the U.S. Department of Agriculture, 1844-1926. (U.S.D.A. Libr. Bibliog. Contrib. 14) 1927. 63 p.

Arranged according to issuing agency or series, i.e., Bureau of Chemistry publications, department bulletins, yearbook articles, etc. Not annotated or indexed.

D32 NATIONAL PHOSPHORUS RESEARCH WORK GROUP. Summary of phosphorus research in the United States relating to soils and fertilizers. [Washington, U.S. National Soil and Fertilizer Research Committee] 1950. 114 p.

Summarizes phosphorus research, largely at the state agricultural experiment stations. Bibliography arranged by subject with an author index, p. 57-114.

D33 NIKLAS, H., ed. Literatursammlung aus dem gesamtgebiet der agrikulturchemie. München, Verlag der Agrikulturchemischen Instituts Weihenstephan der Technischen Hochschule, 1931-1939. 5 v.

Contents: v. 1, Bodenkunde (soil science) by H. Niklas, F. Czibulka, and A. Hock; v. 2, Bodenuntersuchung (soil analysis) by H. Niklas, F. Czibulka, and A. Hock; v. 3, Pflanzenernährung (plant nutrition) by H. Niklas, A. Hock, F. Czibulka, and F. Kohl; v. 4, Düngung und düngemittel (manuring and fertilizers) by H. Niklas, F. Ader, F. Kissel, F. Kohl, and F. Czibulka; v. 5, Ergänzungsband zu band IV (supplement to v. IV) by H. Niklas and F. Ader.

An international classified bibliography covering the whole field of agricultural chemistry. References are not annotated. Indexed by author and subject. Imprint varies. Preliminary material and indexes in German and English.

D34 REITEMEIER, R. F. Bibliography of literature on the release and fixation of native and fixed nonexchangeable potassium and ammonium of soils and silicate minerals. Beltsville, Md., U.S. Bureau of Plant Industry, Soils, and Agricultural Engineering. Division of Soil Management and Irrigation Agriculture, 1952. 28 p.

D35 ROGERS, E. G. and McILVAIN, Z. E. Publications on planning for soil, water, and wildlife conservation, flood control, and land utilization. (U. S. Soil Conservation Serv. Misc. P. 21) 1939. 119 p.

D. Physical Sciences

D36 U.S. BUREAU of CHEMISTRY and SOILS. Index of publications ... 1862-1937. Washington, 1939. 546 p.
V. 1, List of titles and authors, all published. Particularly useful as a finding guide to the Soil Surveys.

D37 U.S. DEPARTMENT of AGRICULTURE. LIBRARY. A classified list of soil publications of the United States and Canada. (Bibliog. Contrib. 13) 1927. 549 p.
Not annotated. Indexed by classes only.

D38 U.S. DEPARTMENT of AGRICULTURE. LIBRARY. Soil conservation bibliographies. Washington, 1940-1942. 4 nos.
No. 1, Wind erosion and sand dune control; a selected list of references. June 1940. 66 p. No. 2, Personnel administration and personnel training; selected list of references. August 1940. 59 p. No. 3, Infiltration of water into soil. October 1940. 76 p. No. 4, Guayule; a list of references. 1942. (Superseded by U.S.D.A. Libr. List 10 (A38)).

D39 U.S. DEPARTMENT of AGRICULTURE. LIBRARY. Soil conservation service library lists. Washington, 1941. 2 nos.
No. 1, Orchard erosion control. 1941. 8 p. No. 2, Economic and social aspects of soil conservation. 1941. 24 p.

D40 U.S. NATIONAL ARCHIVES. Record group 114. Records of Soil Conservation Service; Preliminary check list of records of Soil Conservation Service 1928-1942, comp. by Guy A. Lee and Freeland F. Penney. Washington, 1947. 45 p.

D41 U.S. SOIL CONSERVATION SERVICE. LIBRARY. Soil conservation literature; selected current references. Washington, 1936-1942. 6 v.
Consists of several sections: Periodical articles; Books, pamphlets, state experiment station publications; U.S. government publications; Personnel and training publications. Many of the references are accompanied by brief descriptive annotations. Indexed. Superseded by the Bibliography of agriculture (A24), issued by the U.S. Department of Agriculture Library.

D42 WIELAND, L. H. Bibliography on soil conservation. Rev. by June Henderson, 1936. (U.S. Soil Conserv. Serv. Misc. P. 10) 1936. 179 p.

D43 WIELAND, L. H. Soil erosion bibliography. Washington, Soil Erosion Service, 1935. 124 p.

Dictionaries

D44 GAINES, S. H. Glossary of terms used in soil conservation. (U.S. Soil Conserv. Serv. Misc. P. 15) 1936. 146 p.

D45 A GLOSSARY of special terms used in the soils yearbook. In U.S.D.A. Ybk. of Agr. 1938:1162-1180.
"This glossary is not intended to be a dictionary of all terms used in soil science and related disciplines ... It omits entirely the soil series names and makes no attempt to include the technical terms used in the more specialized phases of soil physics, soil chemistry, and soil microbiology."

D46 INTERNATIONAL SOCIETY of SOIL MECHANICS and FOUNDATION ENGINEERING. Technical terms in English, French, German,

Portuguese, and Spanish used in soil mechanics and foundation engineering. (Sup. to Proceedings of the Third International Conference on Soil Mechanics and Foundation Engineering, Switzerland, 1953.) Zurich, Berichthaus, 1954. 103 p.
Of interest to the agricultural soil scientist since many terms used in soil engineering are also used in agricultural applications.

D47 JACKS, G. V. Multilingual vocabulary of soil science. [Rome] Food and Agriculture Organization of the United Nations [1955?] 439 p.
Based on terms collected by Dr. H. Greene and on reports of national nomenclature committees and individuals in many countries. Arranged by subjects with supplemental lists of terms according to languages.

Directories

See note on directories, p. 15.

D48 COMMERCIAL fertilizer yearbook, 1954. (Commercial fertilizer, v. 89, no. 3-A. September 1954.) Atlanta, Brown, 1954. 177 p.
Issued annually. Contains statistics on fertilizer consumption; list of officials in charge of state fertilizer laws; and a directory of fertilizer manufacturers in the United States, Canada, Cuba, Hawaiian Islands, and Puerto Rico with names of officers. Buyers' directory of fertilizer materials, plant equipment, etc., is also included.

D49 CONSERVATION yearbook. 1952-   Washington, 1952-
Issued annually. Contains information of interest to the soil scientist and conservationist.

D50 FARM chemicals handbook. Ed. 43, 1955/56. Philadelphia, Ware Bros., 1955. 388 p.
Issued annually. Formerly American fertilizer handbook. A directory of fertilizer and pesticide laws and manufacturers. Also includes a "Dictionary of plant foods," and a "Pesticide dictionary."

Handbooks and Manuals

D51 NATIONAL RESEARCH COUNCIL. HIGHWAY RESEARCH BOARD. The use of agricultural soil maps and the status of agricultural soil mapping in the United States. (Highway Res. Board. B. 22) 1949. 128 p.
Important for reference use because of the rated list of soil maps arranged by state and county.

D52 U.S. BUREAU of PLANT INDUSTRY, SOILS, and AGRICULTURAL ENGINEERING. Soil survey manual. (U.S.D.A. Agr. Handb. 18) 1951. 503 p.
A revision of Misc. P. 274. An indispensable reference manual for soil scientists. Includes a brief bibliography on p. 443-454.

D53 U.S. DEPARTMENT of AGRICULTURE. Soils and men. (Ybk. of Agr. 1938) 1938. 1232 p.
Collection of signed articles dealing with problems and causes of

soil misuse, soil management practices, soil and plant relationships, soil science, and the soils of the United States. Glossary of special terms and a reference list of the literature cited in the articles are included. Indexed.

AGRICULTURAL ENGINEERING AND IRRIGATION

The agricultural engineer is fortunate in having good indexes on which to rely. The Engineering index (D54), the "Agricultural engineering" sections in the Bibliography of agriculture (A24) and the Agricultural index (A23) are all useful; the "New books and new bulletins" page of the monthly journal, Agricultural engineering, is helpful as a guide to current literature. The items listed in this section represent only a few of the reference works used in engineering. For a more complete survey see Blanche Dalton's Sources of engineering information (Berkeley, Calif., University of California Press, 1945).

Irrigation is discussed here since it is often thought of as a branch of agricultural engineering. This is misleading, however, since irrigation is also allied to plant physiology and other subjects. The irrigation research worker uses not only general agricultural and engineering bibliographies but also such abstracting journals as Soils and fertilizers (D22), which has a section on reclamation, drainage, and irrigation, and the Bibliography of soil science, fertilizers, and general agronomy (D28).

The diffuseness of irrigation literature is illustrated by the fact that one of the most important recent reviews of research and publication in the field, F. J. Veihmeyer and A. H. Hendrickson's "Soil moisture in relation to plant growth," is in the Annual reviews of plant physiology 1(1950):285-304. Dorothy Graf's bibliography (D59), is one of the few bibliographical compilations concerned with irrigation alone. However, I. E. Hook's Irrigation engineering (New York, Wiley, 1951-   ) and Orson W. Israelson's Irrigation principles and practices (New York, Wiley, 1950) have lists of selected references which are useful.

Abstracting Journals

D54   THE ENGINEERING index. v. 1-   1906-   New York, 1906-
Issued annually. An abstracting publication, arranged by subject, covering the engineering literature of the world. Includes such subjects as agricultural engineering, agricultural machinery, farms, fertilizers, forestry, grain elevators, irrigation, soils, etc. Each volume contains an author index. The publisher also issues a current card service that is superseded annually by the bound volumes.

D55   NATIONAL INSTITUTE of AGRICULTURAL ENGINEERING (Gt. Brit.).
Agricultural and horticultural engineering abstracts. v. 1-   Spring 1950-   Wrest Park, Eng., 1950-
Issued quarterly. A classified abstracting service listing world literature concerning agricultural and horticultural engineering. Each issue has a subject and author index.

Agricultural Engineering

Bibliographies and Indexes

D56 ASSOCIATION of LAND-GRANT COLLEGES and UNIVERSITIES. ENGINEERING DIVISION. Engineering division record. ser. 1- 1921-
Title varies. Formerly <u>Engineering experiment station record</u>. A quarterly bulletin listing by states new publications and research projects of the engineering experiment stations of the land-grant colleges and universities. Lack of general indexes lessens reference value.

D57 DAVIS, E. G. Rural electrification in the United States, a bibliography of selected references. (U.S.D.A. Bibliog. B. 24) 1954. 148 p.
Previously published bibliographies on rural electrification in the United States are listed on p. 1-2. Classified arrangement with author and subject index.

D58 GRAF, D. W. Agricultural engineering; a selected bibliography. Washington, Bureau of Agricultural Engineering, 1937. 373 p.
Lists by subject publications on agricultural engineering of the U.S. Department of Agriculture, the state agricultural experiment stations and extension services from their beginning through 1935. Has an author index.

D59 GRAF, D. W. Irrigation, a selected bibliography. Washington, Bureau of Agricultural Engineering, 1938. 631 p.
Includes references to books, periodicals, and publications of societies through 1937. Arranged by subject. Author index. List of bibliographies, p. 28-29.

D60 INDUSTRIAL arts index, 1913- New York, Wilson, 1913-
Issued monthly with various cumulative issues which are superseded by an annual bound volume. A subject index to a selected list of engineering and technical publications in the English language. Lists items of interest to agricultural engineers under such headings as: Agricultural machinery, Airplanes in agriculture, Electricity on the farm, Rainmaking, etc.

D61 U.S. BUREAU of AGRICULTURAL CHEMISTRY and ENGINEERING. LIBRARY. Current literature in agricultural engineering. v. 1-11, August 1931-March 1942. Washington, 1931-1942. 11 v.
Issued monthly. Title and issuing agency vary. Alphabetical subject index to books, periodical articles, bulletins, etc., received in the library. Some references in each issue are annotated. Not indexed. Superseded by <u>Bibliography of agriculture</u> (A24), prepared by the Library, U.S. Department of Agriculture.

D62 U.S. DEPARTMENT of COMMERCE. BUSINESS and DEFENSE SERVICES ADMINISTRATION. Basic information sources on agricultural machinery and implements. (Business Serv. B. 37) 1954. 7 p.
Lists government and other publications which are useful as sources of information.

D63 WIEBE, R. and NOWAKOWSKA, J. The technical literature of agricultural motor fuels including physical and chemical properties, engine performance, economics, patents, and books. (U.S.D.A. Bibliog. B. 10) 1949. 259 p.
An annotated bibliography arranged chronologically with author and subject indexes and an extensive list of technical publications.

Dictionaries

D64 KAUFFMANN, F. Dictionnaire technique des matériels agricoles et de travaux publics. [Technical dictionary of agricultural and public works machinery.] Français, Anglais, Allemand, Espagnol. [Paris, 1953] 446 p.
 A polyglot dictionary of terms used in connection with agricultural machinery. Subject arranged with alphabetical index for each language.

Directories

See note on directories, p. 15. See Section G, Addendum, for additional material.

D65 AGRICULTURAL engineers' yearbook, 1955. St. Joseph, Mich., American Society of Agricultural Engineers, 1955. 296 p.
 Issued annually. Includes a membership roster, directory of suppliers, professional college curricula, and other information.

D66 FARM implement news' 1954 buyer's guide issue with implement repair information. Chicago, 1954. 416 p.
 Issued annually as pt. 2 of an issue of Farm implement news.

D67 IMPLEMENT and tractor catalog file, 1954. Kansas City, Mo., 1954. 404 p.
 Issued annually. A directory and catalog of farm equipment and supplies.

D68 INTERNATIONAL INSTITUTE of AGRICULTURE. Les institutions de génie rural dans le monde. International directory of agricultural engineering institutions. Ed. 5. Rome, 1939. 152 p.
 French and English. Arranged by continent and alphabetically by country. For each institution information is given regarding location, staff, work in progress, equipment, principal publications issued since 1934, and languages which may be used in correspondence. Lists of organizations, exhibitions, and publications dealing with agricultural engineering are given in a supplementary section. Indexed by name, subject, and country.

D69 MILLARD'S farm equipment directory. v. 64, 1948. Kansas City, Mo., Implement and Tractor, 1948. 237 p.
 Issued annually. Contains an alphabetical list of machinery, tractors, and allied lines with trade names and names and addresses of manufacturers. An additional list gives names of manufacturers together with names of branch offices and distributors.

D70 NATIONAL RETAIL FARM EQUIPMENT ASSOCIATION. Official tractor and farm equipment guide, July-December 1953. St. Louis, Mo., Farm Equipment Retailing, 1953. 184 p.
 Issued twice a year. Provides descriptive information concerning tractors and farm equipment and quotes average values based on sales reports throughout the country.

D71 NATIONAL farm and tractor implement blue book. v. 18, no. 1. Ed. 1957. Chicago, National Market Reports, 1957. 418 p.
 Gives average cash value appraisals, predicated on present-day market conditions, of farm implements and tractors most widely used.

Agricultural Engineering 139

D72 RED tractor book. 39th annual ed., 1955. Kansas City, Mo., Implement and Tractor, 1955. 648 p.
    Issued annually. Contains specifications and operational and dimensional data on all power-operated farm equipment. Official reports of the Nebraska state tractor tests and lists of shop equipment, supply manufacturers, and tractor accessory and equipment manufacturers are also included.

D73 THOMAS' register of American manufacturers. New York, 1954. 5 v.
    Issued annually. V. 1-3, Products classification [list of manufacturers classified by products] A-Z; v. 4, Alphabetical (A-Z) list of manufacturers - trade names - commercial organizations - trade papers; v. 5, Product finding guide to contents and index to advertisers.
    Contains information about manufacturers of agricultural equipment.

D74 TRACTOR field book with power farm equipment specifications. Standard reference book for the design, production, and sale of tractors and power farm equipment, 1953. Chicago, Farm Implement News, 1953. 404 p.
    Issued annually.

D75 WESTERN farm equipment directory, 1956. (Western farm equipment, v. 53, no. 4) San Francisco, 1956. 194 p.
    Issued annually. Lists farm equipment manufacturers and distributors in the western states.

METEOROLOGY

Since the weather is a constant obsession of the agriculturalist, a selected list of meteorological reference works is given here. Winchell's Guide . . . (A8) should be consulted for a more complete listing, including information concerning weather records issued by the U.S. Weather Bureau and other sources. Articles on meterology, particularly from the agricultural viewpoint, can also be found listed in the Agricultural index (A23), the Bibliography of agriculture (A24), and other general indexing sources. Such works as the Atlas of American agriculture (A105), which includes three sections on climate, and C. W. Thornwaite's Atlas of climatic types in the United States (A104), should also be taken into consideration.

Abstracting Journals

D76 AMERICAN METEOROLOGICAL SOCIETY. Meteorological abstracts and bibliography. v. 1- January 1950- Boston, 1950-
    Issued monthly. A continuation in expanded form of the review and abstracting section formerly appearing in the Bulletin of the American Meteorological Society. Arrangement according to Universal decimal classification. Each issue has a current abstracting section and an annotated bibliography. The current abstracting section, which has a portion devoted to agricultural climatology, covers articles published in scientific journals throughout the world. The annotated bibliographies survey the literature of one subject such as soil temperature (March 1951). A "Bibliography of bibliographies on meterology" is in vol. 6, p. 82-115, and an "Annotated

bibliography on textbooks and monographs containing subject bibliographies" is in vol. 6, p. 217-264. Location of items in a few American libraries is indicated. Monthly and annual subject, author, and geographical location indexes.

Bibliographies and Indexes

See Section G, Addendum, for additional material.

D77 HANNAN, A. M. Influence of weather on crops, 1900-1930: a selected and annotated bibliography. (U.S.D.A. Misc. P. 118) 1931. 245 p.
Concerned with influence of weather on crops in connection with the germination, growth, development, susceptibility to diseases, and final yield.

D78 INTERNATIONAL catalogue of scientific literature: F, Meteorology, 1st-14th annual issues. London, Harrison, 1902-1919. 14 v.

D79 ROYAL METEOROLOGICAL SOCIETY, London. Bibliography of meteorological literature ... v. 1- London, 1922-
A semiannual bibliography arranged according to the Universal decimal classification.

D80 U.S. SIGNAL OFFICE. Bibliography of meteorology. Washington, 1889-1891. 4 v.
Subtitle: A classed catalogue of the printed literature of meteorology from the origin of printing to the close of 1881; with a supplement to the close of 1887 and an author index.

Glossaries

D81 GREAT BRITAIN. METEOROLOGICAL OFFICE. The meteorological glossary. Ed. 3. New York, Chemical, 1951. 253 p.
Includes fewer terms than the U.S. Weather Bureau's Weather glossary (D82), but gives more detailed information. Has a list of equivalent terms in Danish, Dutch, French, German, Italian, Norwegian, Portuguese, Spanish, and Swedish.

D82 THIESSEN, A. H. Weather glossary. (U.S. Weather Bur. W. B. 1445) 1946. 299 p.
"This work was designed to meet the general requirements for an American weather glossary. It is not intended as a dictionary but simply defines the words as used in relation to meteorology . . . The references attached to the definitions indicate either sources of quotations or locations of more extended discussions of terms."

Handbooks and Tables

D83 GEIGER, R. The climate near the ground. A translation by Milroy N. Stewart and others of the second German edition of Das Klima der bodennahen Luftschicht. Cambridge, Mass., Harvard University Press, 1950. 482 p.
A definitive work on microclimatology with special reference to the influence of climate on soils and plants. Bibliography, p. 413-462.

D84 SMITHSONIAN INSTITUTION. Smithsonian meteorological tables.
    Ed. 6, prep. by Robert J. List. (Smithsonian Misc. Coll., v. 114.)
    Washington, 1951. 527 p.
D85 U.S. DEPARTMENT of AGRICULTURE. Climate and man. (Ybk. of
    Agr. 1941) 1941. 1248 p.
    A compilation of information and statistics concerning weather,
    with special attention given to agricultural applications. Selected
    lists of references after certain chapters.

SECTION E
FOOD AND NUTRITION

The study of food and nutrition is so closely allied to the study of chemistry, physiology, bacteriology, botany, and medicine that it is difficult to separate its literature from that of these fundamental disciplines. The literature on marketing and sociology of food is also extensive. The items listed here represent only a part of those works specifically concerned with food. Reference should also be made to the Agricultural chemistry, the General agriculture, the Economic botany, and the Commercial fishing and fisheries sections of this compilation. Particularly important works for food and nutrition research which are listed in other sections are: Bibliography of agriculture (A24), which has monthly sections on agricultural products, food, and human nutrition; Chemical abstracts (D8), which has a unit concerning food; and Biological abstracts (A11), which abstracts material on microbiology, food technology, and nutrition.

Since the dairy industry is largely concerned with food products, the majority of the dairy reference works are noted in this section, although the dairy scientists will also use material listed under Animal husbandry.

Some general works on home economics are also listed here since food is a major preoccupation of the home economist. Other titles in this field can be found in Basic books and periodicals in home economics (E4), which was compiled by the Library of the Iowa State College of Agriculture and Mechanic Arts at Ames.

General Works, Reviews, and Bibliographies of Bibliographies

See Section G, Addendum, for additional material.

E1   ADVANCES in food research, ed. by E. M. Mrak and G. F. Stewart. v. 1-   1948-   New York, Academic Press, 1948-
    Issued annually. Reports progress in food research. Contains extensive bibliographies..

E2   AMERICAN CHEMICAL SOCIETY. DIVISION of CHEMICAL LITERATURE. Literature resources for chemical process industries. (Advn. in Chem. no. 10.) Washington, 1954. 582 p.
    Contains a very useful section on the literature of the food industry. Some important articles in this section are: "Introduction and general discussion - food industries literature" by Morris B. Jacobs,

Food and Nutrition 143

"A review of milling and baking literature" by Margaret P. Hilligan and Myrtle J. Krause, "Dairy industry literature" by F. L. Seymour-Jones, "Literature of meat and meat packing" by Barbara J. Payne and H. R. Kraybill, and "Literature of canning and preserving" by Lorraine Ciboch. Each of these articles points out important reference sources, abstracting journals, and bibliographies.

E3 FOOD and AGRICULTURE ORGANIZATION of the UNITED NATIONS. Home economics information exchange. Rome, 1953. 150 p., 8 leaves.

Lists world-wide sources of information concerning home economics organizations and programs, methods and evaluation of teaching, and teaching aids.

E4 IOWA. STATE COLLEGE of AGRICULTURE and MECHANIC ARTS. LIBRARY. Basic books and periodicals in home economics. Ames, Ia., Iowa State College Library, 1942. 112 p.
----- Supplement, 1941-1947. Ames, Ia., Iowa State College Library, 1949. 78 p.

A most useful list compiled chiefly from lists of books on reserve for home economics courses in the libraries of five cooperating institutions. The "Reference books" section is of particular value to bibliographers.

E5 NUTRITION reviews. v. 1- November 1942- New York, Nutrition Foundation, 1942-

Issued monthly. Reviews the literature of nutrition. Annual author and subject indexes.

Abstracting Journals

E6 ABSTRACTS of canning technology. v. 1-8, January 1923-October 1930. Washington, Research Laboratory, National Canners' Association, 1923-1930. 8 v.

Classified bibliography of literature in American and foreign periodicals. Indexed by subject.

E7 AMERICAN CAN COMPANY. Library abstracts. v. 1- 1934- Maywood, Ill., 1934-

Issued monthly. Abstracts food technology literature, placing emphasis on items of interest to the canning industry. List of patents and recent books added to the American Can Company's Library in each issue. Annual author and subject indexes.

E8 AMERICAN DIETETIC ASSOCIATION. Journal. v. 1- June 1925- Baltimore, 1925-

Issued monthly. "Current literature" section abstracts selected literature in the field of dietetics from other American journals, the list of journals varying from month to month. Each volume has a subject index.

E9 AMERICAN SOCIETY of REFRIGERATING ENGINEERS. Refrigeration abstracts. v. 1- January 1946- Easton, Pa., 1946-
Issued monthly. Abstracts over 300 periodicals. Each issue has a section on foods. Has cumulative author and subject indexes. V. 1, no. 2 has a "Subject classification for the literature pertaining to refrigeration and air conditioning."

E. Food and Nutrition

E10 BAKERS' digest. v. 1-   April 1926-   Chicago, 1926-   V. 1-10, no. 5, 1926-April 1936 as Siebel technical review; v. 11-14, no. 6, August 1936-December 1939 as Bakers' technical digest.
Issued bimonthly. Contains abstracts on baking materials and processing, although not in each issue. Has annual author and subject indexes.

E11 BREWERS' digest. v. 9-   1934-   Beloit, Wis., 1934-   Continues in part Siebel technical review, and assumes its numbering. V. 9-12, no. 11, 1934-November 1937 as Brewers' technical review.
Issued monthly. Contains abstracts on brewing materials and processing, although not in every issue. Has annual author and subject indexes.

E12 COMMONWEALTH BUREAU of DAIRY SCIENCE. Dairy science abstracts. v. 1-   May 1939-   Shinfield, Reading, Eng., 1939-
Issued quarterly. Supersedes List of references from current literature, issued by National Institute for Research in Dairying, Reading, Eng. Has review articles and abstracts from world literature. Abstracts are divided as follows: Husbandry, Technology, Control and standards, Economics, Physiology, Bacteriology and mycology, Chemistry and physics, Miscellaneous, Books. There are annual author and subject indexes.

E13 COMMONWEALTH BUREAU of ANIMAL NUTRITION. Nutrition abstracts and reviews, issued under the direction of the Commonwealth Agricultural Bureaux Council, the Medical Research Council, and the Reid Library. v. 1-   October 1931-   Aberdeen, 1931-
Issued quarterly. Classified bibliography of world literature in the field of human and animal nutrition. Sections cover: Techniques, Chemical composition of food stuffs, Vitamins, Physiology of nutrition, Human diet in relation to health and disease, Feeding of animals, and Book reviews. Each issue is indexed by author. Annual author and subject indexes.

E14 FOOD technology. v. 1-   January 1947-   Champaign, Ill., Institute of Food Technologists, 1947-
Issued monthly. Contains book reviews and a "Selected abstracts" section which is occasionally useful for checking and reference. Each volume has a subject and author index.

E15 GENERAL FOODS. RESEARCH and DEVELOPMENT DEPARTMENT. LIBRARY. Current abstracts. v. 1-   1947-   Hoboken, N. J., 1947-
Issued semi-monthly. Abstracts briefly literature concerning food technology and related subjects. Not indexed.

E16 GREAT BRITAIN. DEPARTMENT of SCIENTIFIC and INDUSTRIAL RESEARCH. Food science abstracts. v. 1-   March 1929-   London, 1929-
Issued quarterly. Formerly entitled: Index to the literature of food investigation. Classified, international in scope, and arranged according to kind of food. Also contains sections on processes, packing materials and methods, and engineering. Indexed by authors.

E17 INSTITUTE of BREWING, London. Journal of the Institute of brewing, containing the transactions of the various sections, records of the research scheme, communications, together with abstracts of papers published in other journals, etc. v. 1- March 1895- London, 1895-

Issued bimonthly. Each issue has abstracts of articles concerning brewing and related subjects. There are annual author and subject indexes.

E18　INTERNATIONAL INSTITUTE of REFRIGERATION. Bulletin. v. 1- 1910-　Paris, 1910-

Issued bimonthly. Each bulletin contains extensive abstracts in both French and English from current world literature concerning refrigeration. There are no cumulated author and subject indexes.

E19　INTERNATIONALE zeitschrift für vitaminforschung. International review of vitamin-research. Revue internationale de vitamin-ologie. v. 1-　1932-　Bern, Huber, 1932-

V. 1-18 as Zeitschrift für Vitamin-forschung.

Issued quarterly. Each issue has a lengthy abstracting section. There are annual author and subject indexes.

E20　JOURNAL of dairy science. v. 1-　May 1917-　Baltimore, 1917-

Issued monthly. The section, "Abstracts of literature on milk and milk products," which began with v. 19 (1936), continues Abstract of literature on the production, processing, and distribution of fresh milk and Abstracts of literature on the manufacture and distribution of ice cream. It is an international classified bibliography covering all phases of dairy science, including herd management, feeds and feeding, and genetics and breeding. Annual author and subject index.

E21　JOURNAL of home economics. v. 1-　1909-　Washington, American Home Economics Association, 1909-

Issued monthly. Contains abstracts of literature on such topics as education, family economics, home management, family relations and child development, food, nutrition, housing, institution administration, social welfare and public health, textiles and clothing, etc. Chiefly American literature.

E22　LE LAIT; revue general des questions laitieres. v. 1-　1921- Lyons, Paris, 1921-

Issued bimonthly. Four sections, "Bibliographie analytique," "Bulletin bibliographique," "Brevets," and "Supplement technique," abstract and index world literature on milk and milk products.

E23　MILCHWIRTSCHAFTLICHE forschungen. v. 1-20, 1923-1941. Berlin, Springer, 1923-1941.

"Referatenteil" is a systematic arrangement of abstracts of world literature on milk and milk products and related subjects. Author and subject indexes to each volume.

E24　OFFICE INTERNATIONAL de VIN. Bulletin. Paris, 1938-

Issued monthly. Each issue contains a "Revue de le presse viti-vinicole internationale," which abstracts important articles in world wine periodicals. The "Bibliographie" section abstracts books and monographs.

E25　VITAMIN abstracts. v. 1-　July 1947-　Chicago, Association of Vitamin Chemists, 1947-

Title varies: Abstracts of vitamin literature, 1947-July/September 1949.

Issued quarterly. Abstracts vitamin literature from about 75 journals. Issued in mimeographed form. Reference value is lessened by lack of general indexes.

E26 WALLERSTEIN laboratories communications. v. 1- December 1937- New York, 1937-
Issued quarterly. Has abstracts from world literature of items concerning brewing and related fields. Annual author and subject indexes.

E27 ZEITSCHRIFT für lebensmittel-untersuchung und -forschung. v. 1- 1898- Berlin, 1898-
Issued monthly. A journal of food and nutrition research which includes useful book reviews and a periodical-article abstracting section in each issue. Another special feature is the supplements containing new and world-wide laws and regulations concerning food. Each volume has subject and author indexes.

Bibliographies and Indexes

See Section G, Addendum, for additional material.

E28 AMERINE, M. A. and WHEELER, L. B. A check list of books and pamphlets on grapes and wine and related subjects, 1938-1948. Berkeley, University of California Press, 1951. 240 p.
A comprehensive bibliography for the period covered. The compilers plan to complete a check list of publications from 1900 to 1938 and to continue the above list through five-year supplements.

E29 BITTING, K. Gastronomic bibliography. San Francisco, 1939. 718 p.
Many of the books listed are now in the Bitting Collection, Rare Books Division, Library of Congress.

E30 BRIGGS, C. H. A bibliography of cerealiana. Minneapolis, Miller, 1938. 59 p.
A list of books and pamphlets on cereal production and marketing, milling and baking, and insect and fungus infestations of cereals and cereal products, including publications of state, provincial, and national agricultural experiment stations and boards of agriculture. Compiled for the Northwestern miller.

E31 CHICAGO. UNIVERSITY. INSTITUTE of MEAT PACKING. Books and pamphlets on the meat packing industry. Chicago, 1953. 23 p.
Lists current items issued by the Institute of Meat Packing, the American Meat Institute, and other sources.

E32 CONLEY, A. R. History of research dealing with problems confronting the dairy industry which have been conducted by the agricultural experiment stations, 1881-1947. Bibliography of the research and general index of publications resulting from the research. Columbus, Ohio State University, 1949. 219 p.
Prepared for the Department of Dairy Husbandry and the Department of Rural Economics. The dairy research bulletins of each experiment station are listed chronologically under the name of the state. Subject index.

E33 DYER, MRS. A. I. (ROBERTSON). Guide to the literature of home and family life; a classified bibliography for home economics with use and content annotations. Philadelphia, Lippincott, 1924. 284 p.

Comprehensive bibliography. Includes sections on homemaking, child care and training, food selection, nutrition and dietetics, cookery, interior decoration, handicrafts, and community recreation. Indexed by authors and subjects.

E34   EASTMAN KODAK COMPANY, Rochester, N. Y. DISTILLATION PRODUCTS INDUSTRIES. Annotated bibliography of vitamin E, comp. by Philip L. Harris and Wilma Kujawski of the Research Laboratories. Rochester, N. Y., 1950-1952. 2 v.
Arranged by subject with author index.

E35   ERDMAN, F. S. Bibliography of literature on frozen foods. In Refrigerating Engineering 48(1944):374-380, 414-430.
Covers frozen foods literature issued from 1935 to 1944.

E36   GOURLEY, J. E. Eating round the world, foreign recipe books and magazine articles in English. New York, Printed by Compiler, 1937. 50 p.
A companion list to "Regional American cookery" (E52).

E37   JAHRESBERICHT über die fortschritte in der untersuchung der lebensmittel (Nahrungs- und gemussmittel, sowie gebrauchsgegenstände) ed. by C. A Rojahn. v. 1-46, 1891-1936. Göttingen, 1893-1937. 46 v.
1891-1920, 1922-1936, sonderabdruck a. d. Jahresbericht der Pharmazie. V. 46, 1936 last issued: merged with Jahresbericht der Pharmazie. Bibliography of the international literature on the chemistry of food and food products. Contains sections on milk, butter and margarine, fats and oils, tobacco, coffee, and tea, etc. Author and subject indexes.

E38   LINCOLN, W. American cookery books, 1742-1860, rev. and enl. by Eleanor Lowenstein. Ed. 2. Worcester, Mass., American Antiquarian Society, 1954. 136 p.
Lists about 740 titles arranged by date with indication of libraries holding copies. Author and title indexes.

E39   MANUALE lactis; periodisch erscheinendes handbuch der milchwirtschaftlichen weltliteratur, geordnet nach dem system der Bibliotheca lactis (DK6370 B.L. 90). Gesamt-redaktion: Max E. Schulz und Gerhard Sydow. v. 1-   1948-   Nürnberg, Carl, 1948-
To be completed in 10 volumes. Uses a classification system, "Bibliotheca lactis," devised by Max Erich Schulz. Indexes world dairy literature with emphasis on that published since World War II. Some references include lengthy abstracts.

E40   MERCK & CO., INC. Alpha tocopherol (vitamin E) annotated bibliography. Rahway, N. J., 1941. 92 p.
----- Supplement. Rahway, N. J., 1941. 11 p.
Arranged by subject with an author index.

E41   MERCK & CO., INC. Biotin; annotated bibliography. Rev. June 1944. Rahway, N. J., 1944. 104 p.

E42   MERCK & CO., INC. Nicotinic acid; annotated bibliography, March 1941. Rahway, N. J., 1941. 130 p.
----- Supplement. Rahway, N. J., 1941-

E43   MERCK & CO., INC. Pantothenic acid; annotated bibliography. February 1941. Rahway, N. J., 1941. 50 p.
----- Supplement. Rahway, N. J., 1941-

E44   MERCK & CO., INC. Riboflavin; annotated bibliography. Rahway,

N. J., 1941. 113 p.
----- Supplement. Rahway, N. J., 1941-
E45 MERCK & CO., INC. Vitamin B_1 (thiamine hydrochloride) annotated bibliography. Rahway, N. J., 1941. 140 p.
----- Supplement. Rahway, N. J., 1941-
E46 MERCK & CO., INC. Vitamin B_6; selected annotated bibliography, 1954. Rahway, N. J., 1954. 109 p.
E47 MERCK & CO., INC. Vitamin B_{12}; selected annotated bibliography, 1954. Rahway, N. J., 1954. 245 p.
E48 MERCK & CO., INC. Vitamin K; annotated bibliography. Rahway, N. J., 1941. 78 p.
----- Supplement. Rahway, N. J., 1941. 20 p.
E49 NATIONAL COOPERATIVE PROJECT. Conservation of nutritive value of foods. Beef, an annotated bibliography, comp. by Jessie E. Richardson. (Progress Notes no. 6) Bozeman, Mont., Montana Agricultural Experiment Station, January 1946. 59 p.
E50 NATIONAL COUNCIL on HOTEL and RESTAURANT EDUCATION. Food preparation and related subjects: a selected annotated list of books. Washington, The Council, 1950. 41 p.
E51 GOURLEY, J. E. An annotated list of bibliographies of cookery books. In N. Y. Public Libr. B. 41(1937):696-700.
E52 GOURLEY, J. E. Regional American cookery, 1884-1934; a list of works on the subject. In N. Y. Public Libr. B. 39(1935):452-460, 543-560.

Locates copies in 13 libraries.

E53 STARK, L. M. The Whitney cookery collection, by Lewis M. Stark. In N. Y. Public Libr. B. 50(1946):103-126.

Lists 218 books and manuscripts on cookery, in New York Public Library.

E54 ROTHSCHILD, H. J. N. C. de. Bibliographia lactaria: bibliographie generale des travaux parus sur le lait et sur l'allaitement jusq'en 1899. Paris, Doin, 1901. 584 p.
----- Supplements, 1900-1901. Paris, Doin, 1901-02. 2 v.

Literature about milk from earliest times through 1901. List of milk patents, list of journals. Author index.

E55 SALMONSEN, E. M. Bibliographical survey of vitamins, 1650-1930, with a section on patents by Mark H. Wodlinger. Chicago, Wodlinger, 1932. 334 p.

Arranged chronologically. Most of the references were taken from the John Crerar Library collection. No general indexes.

E56 SCHOELLHORN, F. Bibliographie des brauwesens. Berlin, Gesellschaft für die Geschichte und Bibliographie des Brauwesens, 1928. 436 (i.e. 715) p.
----- Supplements 1-8, 1929-1936.

An international and comprehensive bibliography of the brewing industry arranged in 3 parts: Latin brewing literature, German brewing literature, and Brewing literature in other languages.

E57 SIMON, A. L. Bibliotheca Bacchica, bibliographie raisonnée des ouvrages imprimés avant 1800 et illustrant la soif humaine sous tous ses aspects, chez tous les peuples et dans tous les temps. London, Maggs, 1927-1932. 2 v.

Contents: t. 1, Incunables; t. 2, Seizième siècle.

E58 SIMON, A. L. Bibliotheca gastronomica, a catalogue of books and

Food and Nutrition

documents on gastronomy. London, Wine and Food Society, 1953. 196 p.

Largely a catalog of rare and interesting books on wine and food from the author's private collection. Arranged alphabetically by author with short title and subject indexes.

E59 SIMON, A. L. Bibliotheca vinaria; bibliography of books and pamphlets dealing with viticulture, wine-making, distillation, the management, sale, taxation, use and abuse of wines and spirits. London, Richards, 1913. 339 p.

"In the present bibliography are included not only all the books which form the library of the Wine Trade Club, but also all those which it is hoped to add to the club collection."

E60 STECHOW, M. Register der weltliteratur über vitamine ... v. 1- 1943-
V. 1 (Leipzig, Holingsche Verlagsanstalt, 1943) covers vitamin literature from 1890 to 1929; V. 2 (Würzburg, Physica-Verlag, 1956) covers 1930-1945. Arranged by broad subjects, has author indexes and lists of periodicals.

E61 STROBEL, D. R., BRYAN, W. G., and BABCOCK, C. J. Flavors of milk, a review of literature. Washington, U.S. Department of Agriculture, 1953. 91 p.

A literature review with an alphabetically arranged author list of references cited.

E62 U.S. AGRICULTURAL RESEARCH SERVICE. STATE EXPERIMENT STATIONS DIVISION. Published and processed reports of research in foods, human nutrition, and home economics at the land-grant institutions, 1935/36- Washington, 1936-

Issued annually. Title varies: 1935/36-1942/43, Research in home economics at the land-grant institutions; 1944/45-1949/50, Research in foods, human nutrition, and home economics at the land-grant institutions. Arranged by states and then classified by subjects. No cumulative author or subject indexes. From 1935/36 to 1945/46 published works and research in progress were included; since 1946/47 only publications are listed.

E63 U.S. BUREAU of HUMAN NUTRITION and HOME ECONOMICS. Titles of completed theses in home economics and related fields in colleges and universities of the United States, 1942/46- Washington, 1946-

Issued annually. Title varies: Completed theses in home economics and related fields in colleges and universities of the United States, 1942/46-1947/48.

Arranged by broad subjects with author indexes. Has been issued in cooperation with the Office of Education since 1948/49.

E64 U.S. DEPARTMENT of AGRICULTURE. LIBRARY. Bibliography on dehydration of foods, 1938-1943. (U.S.D.A. Bibliog. B. 6) 1945. 120 p.

E65 U.S. DEPARTMENT of COMMERCE. INQUIRY REFERENCE SERVICE. Food processing, marketing, and equipment (Basic information sources). Washington, January 1949. 10 p.

Lists government publications and other sources containing information on this subject. Also lists directories, trade papers, and associations. See also Grocery wholesaling (1949); Food stores (1947); Grocery, meat, and produce stores (1950); and Frozen foods and lockers (1948) issued by the same department.

150   E. Food and Nutrition

E66   U. S. OFFICE of EDUCATION. Notes on research in home economics education, 1934/36-1942/45. Washington, 1936-1946. 8 nos.
Title and issuing body varies. Includes abstracts of doctoral dissertations and professional studies in home economics education. Partly continued by E63.

E67   U. S. OFFICE of TECHNICAL SERVICES. Bibliography of reports on foods and the food industry, April 1949. Washington, 1949. 52 p.
Lists items abstracted in its Bibliography of scientific and industrial reports.

E68   U. S. QUARTERMASTER FOOD and CONTAINER INSTITUTE, Chicago, Ill. Bibliography on the deterioration of fat in foods. [1946?] Issued on 3,456 index cards. Lists and abstracts publications on deterioration of fat in foods. Prepared by Dr. B. F. Daubert and staff of the University of Pittsburgh. Distributed to various technical libraries in the United States. Index . . ., by B. F. Daubert and Marjorie Benoy (44 p.), was issued in April 1949.

E69   U. S. QUARTERMASTER FOOD and CONTAINER INSTITUTE, Chicago, Ill. LIBRARY BRANCH. Bibliographic series. no. 1. Chicago, 1953-
Entries arranged by author. Subject indexes. Retrospective bibliographies covering world literature relating to food.

E70   VICAIRE, G. Bibliographie gastronomique. La cuisine.- La table.- L'office.- Les aliments.- Les vins.- Les cuisiniers et les cuisinières.- Les gourmands et les gastronomes.- L'économie domestique.- Facéties.- Dissertations singulières.- Pièces de théâtre, etc. etc., dupuis le xve siècle jusqu'à nos jours. Avec des fac-similés. Paris, Rouguette, 1890. 972 col. [Reprint ed., London, Verschoyle, 1954]
An old but extensive bibliography still useful historically.

E71   WEIL, B. H. and STERNE, F. Literature search on the preservation of food by freezing; jointly sponsored by the State engineering experiment station and the Tennessee Valley authority. (Spec. Rpt. no. 23.) Atlanta, State Engineering Experiment Station, 1946. 409 p.

-----   First Supplement, Jan. 1946-July 1947, by Betty Anderson and B. H. Weil. In Special Rpt. 25(1948):407-670.

E72   WINE INSTITUTE, San Francisco. Selective bibliography of wine books ... rev. to Feb. 1, 1947. San Francisco, 1947. 39 p.
A mimographed list arranged by broad subjects.

Dictionaries and Encyclopedias

See Section G, Addendum, for additional material.

E73   BLÜHER, P. M. Meisterwerk der speisen und getränke; Französisch-Deutsch-Englisch (und andere sprachen). Nordhausen am Harz, Killinger, 1928. 2 v.
An international encyclopedia of food and drink.

E74   BRUNET, R. Dictionnaire d'oenologie et de viticulture. Paris, Ponsot, 1946. 535 p.
Pt. 1 is a dictionary of terms used in the wine industry. Pt. 2 supplies information and maps of the wine producing regions of

France. Pt. 3 is a buyer's guide to equipment used in the French wine industry.

E75 DAHL, J. O. Dictionary of 1001 menu terms; foods, wines, spirits, cocktails. Stamford, Conn., The Author, 1936. 47 p.

E76 DAIRY INDUSTRIES SOCIETY, INTERNATIONAL. French and English glossary of dairy and related terms. Washington, 1954. 64 p.

E76a DAIRY INDUSTRY SOCIETY, INTERNATIONAL. Spanish and English glossary of dairy and related terms. Washington, 1949. 47 p.

E77 DAVIS, J. G. A dictionary of dairying. Ed. 2. New York, Interscience, 1955. 1132 p.

An encyclopedic compendium of basic information concerning dairying by a British scientist. Includes a number of lengthy articles by specialists.

E78 FAES, H. Lexique viti-vinicole international. Français, Italien, Espagnol, Allemand. Paris, Office International du Vin, 1940. 278 p.

Lists French, Italian, Spanish, and German wine terms in current usage.

E79 FISCHER, W., and others. Fachwörterbuch für brauerei und mälzerei. Nürnberg, Carl, 1946. 71 p.

Dictionary of terms used in the brewing industry.

E80 FOOD industries manual; a technical and commercial compendium on the manufacture, preserving, packing, and storage of all food products. [Ed. 15] London, Hill, 1947. 1168 p.

In encyclopedic form. Contains the following sections: Cereals, Sugar confectionery, Canning and preserving, Meat products, Pickles and sauces, The dairy industry, Food dehydration, Edible fats and fatty foods, Fruit juice products, Storage and refrigeration, Handling equipment, Packaging, Boiler house, and Composition of foods. Subject index and list of trade names and marks are also included.

E81 GARRETT, T. F., ed. The encyclopaedia of practical cookery; a complete dictionary of all pertaining to the art of cookery and table service. London, Gill, 1898. 4 v.

Includes, in addition to recipes, descriptions of utensils and explanations of culinary terms and processes. Illustrated, alphabetically arranged, and indexed.

E82 GRAVES, L. G. and TABER, C. W. A dictionary of food and nutrition. Philadelphia, Davis, 1938. 423 p.

Compendium of terms related to the science of nutrition and dietetics. Tables showing baking and refrigeration temperatures, normal height and weight for age, and nutritive value of foods are given in an appendix.

E83 KINGAN & CO., INC. A dictionary of packinghouse terms. Indianapolis, 1951. 101 p.

E84 KRACÍKOVÁ, L. Rusko-český slovník z oboru masného prumyslu. Vyd. 1. Praha, Státní Nakl. Technické Literatury, 1953. 112 p.

Russian-Czech meat dictionary.

E85 KNÖCHEL, E. Fachwörterbuch für die zuckerindustrie; Deutsch-Italienisch und Italienisch-Deutsch. Berlin, Instituts für Zucker-Industrie, 1936. 142 p.

Dictionary of technical terms used in the sugar trade.

E. Food and Nutrition

E86   KRAMER, O. Kellerwirtschaftliches lexikon. Neustadt an der Weinstrasse, Meininger, 1954. 276 p.
     A wine lexicon which also contains information concerning German wine laws and regulations, the wine industry in Germany and foreign countries, and a list of German wine equipment suppliers.

E87   MARTINY, B. Wörterbuch der milchwirtschaft aller länder. Leipzig, N. Heinsius Nachfolger, 1907. 142 p.
     A dictionary of dairying terms used in several languages.

E87   MÜLLER, K. Weinbau-lexikon. Berlin, Parey, 1930. 1015 p.
     An encyclopaedia of oenology and viticulture.

E88   SANZ EGAÑA, C. Enciclopedia de la carne; producción, comercio, industria, higiene. Madrid, Espasa-Calpe, 1948. 966 p.
     A Spanish encyclopedia on meat and meat industry.

E89   SIMON, A. L., ed. A concise encyclopaedia of gastronomy. London, Wine and Food Society, 1939-1946. 9 pts. London, Collins, 1952. 827 p. New York, Harcourt, Brace, 1952. 816 p.
     Contents: Sect. 1, Sauces; sect. II, Fish; sect. III, Vegetables; sect. IV, Cereals; sect. V, Fruit; sect. VI, Birds and their eggs; sect. VII, Meat; sect. VIII, Wines and spirits; sect. IX, Cheese and analytical cross-index to all sections.
     Contains recipes and definitions of terms.

E90   SIMON, A. L. A dictionary of wine. New York, Longmans, Green, 1936. 266 p.

E91   SCHOONMAKER, F. Frank Schoonmaker's dictionary of wines, ed. by Tom Marvel. New York, Hastings House, 1951. 120 p.
     An alphabetical list of wines, wine districts, and wine terminology. Also includes a dictionary of alcoholic beverages other than wine.

E92   SMITH, H. The master dictionary of food and cookery and menu translator. New York, Philosophical Library, 1950. 263 p.

E93   WARD, A. The encyclopedia of food. The stories of the foods by which we live, how and where they grow and are marketed, their comparative values, and how best to use and enjoy them. New York, The Author, 1923. 596 p.
     Well-illustrated, nontechnical descriptions and articles. A dictionary of food names in six languages - English, French, German, Italian, Spanish, and Swedish - and a list of culinary and bill-of-fare terms are given in the appendix.

Directories

See note on directories, p. 15.

E94   ADRESSBUCH der Deutschen nahrungs- und genussmittelbetriebe. Industrie und grosshandel. Berlin, Kaupertverlag, Deutsche Adressbuch-Gesellschaft, 1950. 1178 p.
     Lists food industry firms in all parts of Germany, alphabetically under region and by speciality. There are also lists of food organizations and publications and manufacturers of machinery used in food production.

E95   AMERICAN beet sugar companies, 1953-1954. Washington, United States Beet Sugar Association, 1954. [12] p.
     Has statistical information about sugar beets in the United States

Food and Nutrition 153

and lists American sugar beet companies.

E96 AMERICAN brewer register, 1957. The brewery buyer's guide. (American brewer, December 1956, pt. II.) New York, American Brewer, 1956. 66 p.
Issued annually. A directory of products used by brewers. Also has a company address and a trade names section.

E97 AMERICAN MEAT INSTITUTE, Chicago. DEPARTMENT of PUBLIC RELATIONS. Reference book of the meat industry. Ed. 1. 1922-   Chicago, 1922-
Issued irregularly.

E98 ANNUAL meat packers' guide, 1955. Chicago, National Provisioner, 1955. 232, 124 p.
Has a "Management section" listing trade associations, production and consumption data, general food laws, state and federal regulations, and government agencies; a "Construction section" noting information on federal inspection requirements, meat plant construction and design, packinghouse department layouts and containing a two page list of publications and literature of interest to the meat packing industry. There is also an extensive classified buyers' guide to equipment and supplies.

E99 ANNUARIO vinicolo d'Italia. Ed. 23, 1951-1952. Milan, Unione Italiana Vini [1951?] 760 p.
A directory of the Italian wine industry.

E100 BAKING industry. Bakers' buying directory issue, 1954. Chicago, Clissold, 1954. Variously paged.
Issued annually. Sources of machinery, equipment, supplies, and services for the baking industry. Technical reference section and trade names index are also included.

E101 BREWERS' digest annual buyers' guide and directory, 1955. Chicago, Brewers' Digest, 1955. 170 p.

E102 CANADA food directory, 1946-1947. Ottawa, Canadian Trends, 1947. 131 p.
Lists names and addresses of brokers, wholesalers, and chain and department stores, by provinces and cities; shippers, packers, canners, and processors, classified by product; cold storage plants; box and container manufacturers. Includes a statistical section showing acreage and production of fruits, vegetables, and field crops, and imports and exports, food items. Commodity index.

E103 CANADIAN dairy and ice cream journal. Dairy industry buyers' directory. Toronto, Smith, 1956. 108 p.
Issued annually.

E104 CANDY industry catalog and formula book, 1954/55. New York, Candy Industry, 1954. 256 p.
Issued annually.

E105 DAIRY industries catalog of equipment, supplies, and services used by dairy products manufacturers. Ed. 28. Milwaukee, Olsen, 1955. 492, 65 p.
Issued annually. Directory of manufacturers of dairy supplies and equipment arranged alphabetically by product. Also included are lists of trade names; dairy industry associations; workers in dairy schools and experiment stations; state health, food, and drug officials; tables of industry statistics; and tables of weights, measures, conversion factors, etc.

154 E. Food and Nutrition

E106 DEUTSCHLANDS weinbauorte und weinbergslagen, bearbeitet von Dr. Eduard Goldschmidt. Ed. 6. Mainz, Verlag der Deutschen Wein-Zeitung [1951] 263 p.
   Issued irregularly. A directory of the German wine industry, arranged geographically.

E107 FOOD industries catalogs for food processors. Ed. 11, 1949-1950. New York, McGraw-Hill, 1949. Various paging.
   Issued annually. Collection of catalogs of manufacturers of machinery, equipment, supplies, and ingredients for food processors, with product, company name, and trade name indexes.

E108 FOOD PACKER. Annual buyers' guide and reference issue, 1954 (Food packer, Oct. 15, 1954). Pontiac, Ill., 1954. 112 p.

E109 FROZEN food factbook and directory, 1956. New York, National Wholesale Frozen Food Distributors Association, 1956. 156 p.
   Has statistics and other information concerning the frozen food industry and directories of committees, officials, organizations, distributors, refrigerated warehouses, suppliers, etc.

E110 HARPER'S directory and manual, 1952: the standard work of reference for the wine and spirit trade. London, Harper, 1952. 590 p.
   A directory of the British wine industry.

E111 INSTITUTE of FOOD TECHNOLOGISTS. Directory of members, 1956. Chicago, 1956. 128 p.
   Issued irregularly.

E112 THE INTERNATIONAL green book of cottonseed and other vegetable oil products. v. 39, 1953-1954. Dallas, Cotton Gin and Oil Mill Press, 1953. 576 p.
   Issued annually. Directory containing lists of officers and committees of the National Cottonseed Products Association, Inc.; oil mills in the United States and foreign countries; mixed feed and fertilizer manufacturers in the United States; and cotton compresses in the United States.

E113 MANUAL of sugar companies, 1952/53. Ed. 30. New York, Farr, 1953. 324 p.
   Issued annually. Contains statistics and other information and lists sugar companies of North America and the West Indies.

E114 MEAT magazine. Meat packing cyclopedia. Chicago, 1950. 915 p.
   Largely a geographical list of meat packers in the United States and a buyers' guide.

E115 MODERN brewery age blue book issue, 1954. Chicago, Brewery Age, 1954. 242 p.
   Issued annually. Contains industry information and directories of the Western Hemisphere.

E116 NATIONAL CANNERS' ASSOCIATION. Canners' directory, 1954: with indications of membership in the National canners' association and lists of members of the Canning machinery and supplies association, the National food brokers' association. Washington, 1954. 253 p.
   Issued annually.

E117 THE NORTHWESTERN miller. Almanack number, April 28, 1953. Minneapolis, Miller, 1953. 144 p.
   Issued annually. Contains legislation, statistics, and other information concerning the milling industry. Has list of trade associations and a selected bibliography, p. 42-44, on milling, baking, wheat, and feed.

Food and Nutrition 155

E118 THE NORTHWESTERN miller. A list of flour mills in the United States and Canada, with supplementary list of blending plants, dry corn mills, rye mills, durum mills, and buckwheat mills. [Minneapolis] Miller, 1951. 149 p.

E119 PRODUCE REPORTER COMPANY. Semi-annual blue book. A credit and marketing guide adopted to the daily business needs of wholesale handlers and users of perishable fresh fruits and vegetables in the United States and Canada, April 1957. Wheaton, Ill., 1957. 768, S191 p.
Information about handlers of fruits, vegetables, and wholesale groceries listed geographically and by commodity. Also has information about regulatory measures, laws, contract forms, rules and customs, etc.

E120 QUICK frozen foods (periodical). 1954 directory of frozen food processors of fruits, vegetables, seafoods, meats, poultry, concentrates, and prepared foods. New York, Williams, 1954. 768 p.
Issued annually. Includes foreign countries as well as the United States.

E121 RED book, encyclopaedic directory of the wine and liquor industries, 1953. New York, Liquor Publications, 1953. 360 p.
Issued annually. Contains directory of United States wine and liquor producers, United States brands, imported brands, state administrative officials, and equipment and supplies. Information concerning state and federal laws and regulations is also supplied.

E122 LA SEMANA vitivinicola. Directorio de la produccion, comercio y exportacion de vinos, alcoholes, vinagres, licores y de las industrias conexas, asi como de las principales casas extranjeras importadoras de vinos y espirituosos, 1952/53. Valencia, 1952.
A directory of the Spanish wine trade.

E123 SOYBEAN blue book, 1954. Hudson, Ia., American Soybean Association, 1954. 160 p.
Issued annually. Contains lists of officers and information about the American Soybean Association, National Soybean Processors' Association, Soy Flour Association; official standards for soybeans; list of definitions and product descriptions; tables showing production, utilization, and prices of soybeans and soybean products; list of processors, refiners, and manufacturers; and an industry buyers' guide.

E124 SUGAR reference book, 1957. Ed. by E. W. Mayo, Jr. New York, Palmer, 1957. 198 p.
Issued annually. Contains the year's developments, statistical data, and a directory of equipment and manufacturers.

E125 THOMAS' wholesale grocery and kindred trades register. 56th, 1954. New York, Thomas, 1954. 1818 p.
Issued annually. Contains lists of wholesale grocers, brokers, chain stores in the United States and Canada and a list alphabetically arranged by products of manufacturers, packers, etc., of canned, frozen, and dried foods, flour, feed, meal, insecticides, etc. Has index of company names.

E126 UKERS' international tea and coffee buyers' guide. Ed. 17, 1954/55. New York, Tea and Coffee Trade Journal, 1954. 448 p.
Issued biennially. Has buying guide, descriptive and statistical

E. Food and Nutrition

information concerning tea and coffee throughout the world, tea and coffee dictionaries, list of associations and organizations, etc.

E127 U.S. AGRICULTURAL MARKETING SERVICE. Wholesale meat trade directory. v. XXIII. Washington, 1954. 45 p.
Issued irregularly.

E128 UNITED States food products directory, the blue book of food packers and distributors. v. XIII, 1955. San Francisco, Western Canner and Packer, 1955. 546 p.
Issued irregularly. United States food canners and packers are listed alphabetically by state and in a classified section by products. Food distributors and brand names are also noted in special lists. A buyers' guide is provided.

E129 U.S. PRODUCTION and MARKETING ADMINISTRATION. Directory of refrigerated storage warehouses in the United States. Washington, 1953. 92 p.
Issued irregularly. Lists the names, addresses, and type of operation of all refrigerated storage warehouses currently cooperating with the U.S. Department of Agriculture in the issuance of the Monthly cold storage report and the Biennial space survey.

E130 WESTERN canner and packer. 1952. Statistical review and yearbook number. (Western Canner and Packer, May 25, 1952.) San Francisco, Miller Freeman Publications, 1952. 222 p.
Issued annually. Contains a review of the processed foods industry for the year together with tables showing production of canned, frozen, and dehydrated fruits and vegetables, etc. A list of industry associations is included.

E131 WHO'S who in meat, a directory of the meat industry. New York, New York Council of Wholesale Meat Dealers, 1954. 132 p.

E132 WHO'S who in the dairy industries, 1953. New York, Urner-Barry, 1953. 458 p.
Issued annually. Lists dealers in equipment and supplies; producers, shippers, wholesale receivers, distributors, and brokers of butter, cheese, milk, and ice cream products. Has a butter brand name index. Information concerning federal agencies, associations, and organizations is also listed.

E133 WHO'S who in the egg and poultry industries, 1954. New York, Urner-Barry, 1954. 442 p.
Issued annually. Directory of equipment suppliers, packers, shippers, hatcheries, wholesale receivers and brokers, trade organizations, state and county officials, warehouses, etc. Also has general and statistical information about the industry.

E134 WINES and vines. Annual directory issue, 1954. (Wines and vines, September 15, 1954.) San Francisco, 1954. 148 p.
Contains directory of wineries, United States wine brand index, buyers' guide, and state laws and regulations.

Handbooks, Manuals, etc.

E135 AMERICAN CAN COMPANY. The canned food reference manual. Ed. 3. New York, 1947. 638 p.
Contains various useful tables and data concerning canned and fresh foods. "Selected bibliography on canned foods," p. 597-638.

E136 AMERICAN HOME ECONOMICS ASSOCIATION. FOOD and NUTRITION DIVISION. TERMINOLOGY COMMITTEE. Handbook of food preparation. Washington, 1954. 66 p.
A useful handbook of facts and figures concerning food. Has information on weights and measures, recipe construction, timetables, definitions of foods, properties of foods, and quality and size grades.

E137 AMERICAN MEDICAL ASSOCIATION. COUNCIL on FOODS and NUTRITION. Handbook of nutrition. Ed. 2. New York, Blakiston, 1951. 717 p.
An authoritative compilation of food and nutrition data with bibliographical notes.

E138 BLANCK, F. C., ed. Handbook of food and agriculture. New York, Reinhold, 1955. 1039 p.
A review by specialists of the present state of knowledge in food and agriculture. Has tables, statistics, and selected references at end of chapters. Appendix contains important United States food laws, nutritional tables, and information about special food agencies and research groups. Also included is a useful "Guide to the literature on food research and technology," p. 974-989.

E139 THE CANNING trade almanac. 1954, 39th annual compilation. Baltimore, Canning Trade, 1954. 372 p.
Issued annually. Contains list of industry associations with principal officers; text of the Federal Food, Drug, and Cosmetic Act; food laws and regulations; United States canned food grades; labeling requirements; and industry statistics. List of members of the National Food Brokers' Association is also included.

E140 JACOBS, M. B. The chemical analysis of foods and food products. Ed. 2. New York, Van Nostrand, 1951. 902 p.
An important standard reference manual supplying basic information concerning the analysis and composition of foods.

E141 JACOBS, M. B., ed. The chemistry and technology of food and food products. Ed. 2. New York, Interscience, 1951. 3 v.
Signed articles, contributed by specialists, covering the production, chemistry, sanitation, and control of foods. List of contributors and subject indexes are included.

E142 MCLESTER, J. S. Nutrition and diet in health and disease. Ed. 5. Philadelphia, Saunders, 1949. 800 p.
The appendix has useful tables of food data. Bibliographical references are supplied at the end of chapters.

E143 MATTICE, M. R. Bridges' food and beverage analyses. Ed. 3. Philadelphia, Lea & Febiger, 1950. 412 p.
A most useful manual with numerous tables providing data concerning food and its constituents. Has a glossary of vitamin terms and a bibliography.

E144 NATIONAL RESEARCH COUNCIL. COMMITTEE on FOOD HABITS. Manual for the study of food habits. (Natl. Res. Council. B. 111) Washington, 1945. 142 p.
Contains a selected but lengthy bibliography, p. 101-135.

E145 NATIONAL RESEARCH COUNCIL. COMMITTEE on SURVEY of FOOD and NUTRITION RESEARCH. Survey of food and nutrition research in the United States. 1947-   Washington, 1948-
Issued irregularly. Compiled with the support of the Committee on

Food Research of the Quartermaster Food and Container Institute for the Armed Forces. Lists research projects by subject. Also includes directories of organizations and personnel.

E146 OFFICE INTERNATIONAL du VIN. Annuaire, 1951. Paris, 1951. 890 p.
Contains legislation concerning wine in effect in the major countries of the world, statistics, wine associations, lists of world wine journals, and other information.

E147 PARRY, J. W. The spice handbook; spices, aromatic seeds, and herbs. Brooklyn, Chemical, 1945. 254 p.
In 5 parts: pt. 1, Extracts from the pure food laws and regulations of the United States and Canada; pt. 2, Spices; pt. 3, Aromatic seeds; pt. 4, Herbs; and pt. 5, Spice formulae.
Under each commodity information is given regarding the plant, family, nativity and cultivation, properties and uses, and government standards, etc. Standards contracts of the American Spice Trade Association are given in an appendix. Subject index.

E148 WINTON, A. L. and K. B. The structure and composition of foods. New York, Wiley, 1932-1939. 4 v.
V. 1, Cereals, starch, oil seeds, nuts, oils, forage plants; v. 2, Vegetables, legumes, fruits; v. 3, Milk (including human), butter, cheese, ice cream, eggs, meat extracts, gelatin, animal fats, poultry, fish, shellfish; v. 4, Sugar, sirup, honey, tea, coffee, cocoa, spices, extracts, yeast, baking powder. Each volume indexed by subject.

Periodical Lists

See Section G, Addendum, for additional material.

E149 GUILLETMOT, A. Förteckning över periodiska publikationer rörande mejerihanteringen. In Nordisk Jordbrugsforskning 35(1953): 117-177.
World list of periodicals on dairy science.

Tables

E150 BOWES, A. De P. and CHURCH, C. F. Food values of portions commonly used. Ed. 6. Philadelphia, The Author, 1946. 58 p.
Includes bibliographies.

E151 BRADLEY, A. V. Tables of food values. Peoria, Ill., Manual Arts Press, 1942. 224 p.
Contains extensive tables showing the food value of average servings or of convenient measures of commonly used foods. Bibliography, p. 216-220.

E152 FOOD and AGRICULTURE ORGANIZATION of the UNITED NATIONS. NUTRITION DIVISION. Food composition tables for international use, by Charlotte Chatfield. (F.A.O. Nutritional Studies, no. 3.) Washington, 1949. 56 p.
The bibliography of source materials lists 158 items.

E153 FOOD and AGRICULTURE ORGANIZATION of the UNITED NATIONS. NUTRITION DIVISION. Food composition tables, minerals, and vitamins for international use, by Charlotte Chatfield (F.A.O. Nutritional Studies, no. 11.) Rome, 1954. 117 p.

"This publication is in sequence with Food composition tables for international use (E152) published by F.A.O. in 1949. The earlier tables showed the calorie value and the protein, fat, and carbohydrate content of foods commonly used throughout the world. The present tables give the figures for vitamin A, ascorbic acid (vitamin C), thiamine (vitamin B_1), riboflavin and niacin, and for two minerals, calcium and iron." Bibliography, p. 59-117.

E154 WOOSTER, H. A. Nutritional data. Ed. 2. Pittsburgh, H. J. Heinz Co., 1954. 155 p.

Tables provide data on proximate composition, minerals, vitamins, and average portions of common foods.

E155 HEWSTON, E. M., and others. Vitamin and mineral content of certain foods as affected by home preparation. (U.S.D.A. Misc. P. 628) 1948. 76 p.

Contains vitamin and mineral content tables of cooked foods. Literature cited, p. 65-68.

E156 NATIONAL RESEARCH COUNCIL. FOOD and NUTRITION BOARD. Tables of food composition giving proximate mineral and vitamin components of foods (on a 100 gram basis) based on N. R. C. November 1, 1943 Army tables. Washington, 1944. 22 p.

The tables show nutritive value of 17 groups of foods, food products, and beverages.

E157 TAYLOR, C. M. Food values in shares and weights. Reissue. New York, Macmillan, 1954. 190 p.

Contains tables giving food values of common measures of common foods.

E158 U.S. BUREAU of HUMAN NUTRITION and HOME ECONOMICS. Tables of food composition in terms of eleven nutrients. (U.S.D.A. Misc. P. 572) 1945. 30 p.

Average values for food energy, protein, fat, carbohydrate, three minerals, and the better known vitamins are given for a selected list of foods.

SECTION F
SOCIAL SCIENCES

AGRICULTURAL ECONOMICS, STATISTICS, AND LEGISLATION

The agricultural economist is actually a social scientist working with agricultural problems. The literature he uses is widespread and includes publications in general economics, statistics, finance, political science, and in history; in fact, all of the social sciences. The best bibliography of this material, as it applies to agriculture, is Miss Orpha E. Cummings' Important sources of information for work in agricultural economics with special emphasis on California (F1). Although its emphasis is on California, this most excellent compilation describes the major reference sources and bibliographical tools in agricultural economics and is of primary importance to the researcher or librarian in this field. Because it is such a comprehensive guide, the list given here has been restricted to the most important items. Readers are also reminded that the Agricultural index (A23) indexes part of the literature of agricultural economics published in this country and that the comprehensive Bibliography of agriculture (A24) has a monthly section entitled "Agricultural economics and rural sociology" which lists world literature. Since publications of the U. S. Department of Agriculture's Agricultural Marketing Service (which includes the major part of the former Bureau of Agricultural Economics and the Production and Marketing Administration) are so important, careful attention should also be given to Agricultural economic and statistical publications (F7) and Periodic reports (F5). Economic publications of the U. S. Department of Agriculture are also indexed in the general indexes to U. S. Department of Agriculture publications noted in the "Agriculture - General" section.

Those wishing guidance in the use of the general literature of economics and the social sciences will wish to refer, among other works, to Louis Kaplan's Research material in the social sciences (Madison, Wis., University of Wisconsin Press, 1939), Winchell's Guide to reference books (A8), and Edwin T. Coman's Sources of business information (New York, Prentice Hall, 1949). One should also keep in mind that the Encyclopedia of the social sciences, although dated, is still of great value.

General Works

F1 GIANNINI FOUNDATION of AGRICULTURAL ECONOMICS. LIBRARY.
    Important sources of information for work in agricultural

economics with special emphasis on California [by Orpha E. Cummings]. Ed. 9. Berkeley, Calif., 1956. 108 p.

Guide to reference books. Contains lists of general and special bibliographies, directories, handbooks, journals, statistical sources, both foreign and domestic, and agencies which issue statistics as well as indexes relating to economic conditions. Classified and annotated. Not indexed.

Bibliographies and Indexes

F2   FRAUENDORFER, S. VON. Internationale bibliographie der agrarökonomischen schrifttums. In Bearb. in der Bibliothek des Internationalen Landwirtschafts-Institut. Berichte über Landwirtschaft ... n. f. 17(1933):726-768. 24(1939):729-777.
Classified bibliography of the world's literature in the field of agricultural economics, including land settlement, credit, cooperation, insurance, trade, prices, taxation, statistics, farm management, labor, agricultural history and geography, education, and rural sociology. Continued by Bibliographie internationale d'économie rurale (F3).

F3   INTERNATIONAL INSTITUTE of AGRICULTURE. Bibliographie internationale d'économie rurale. International bibliography of agricultural economics. Internationale bibliographie der agrarwirtschaft. v. 1-8, October 1938-December 1946. Rome, 1938-1946. 8 v.
French, English, and German. Compiled from publications received currently in the Library of the International Institute of Agriculture. Indexes books, pamphlets, periodical articles, etc., in all languages "dealing with the economic and social aspects of agriculture in the widest sense of the words." Titles only. Classified. Annual author indexes. Continues International Bibliographie der Agrarökonomischen Schrifttums (F2).

F4   U.S. AGRICULTURAL MARKETING SERVICE. Check list of reports issued by Agricultural Marketing Service. Washington, 1941-
A monthly list supplementing the U.S. Bureau of Agricultural Economics' Available printed and processed publications issued from July 1, 1940 to February 28, 1941 (F8). Title from 1941 to 1953 was Check list of printed and processed publications of the Bureau of Agricultural Economics; from 1953 - 1955 title was Check list of reports and charts.

F5   U.S. AGRICULTURAL MARKETING SERVICE. Periodic reports. (AMS-48) 1955. 53 p.
Lists market reports issued by the Agricultural Marketing Service which has supplanted, in part, the former Production and Marketing Administration. Also lists offices at which market reports are issued.

F6   U.S. AGRICULTURAL RESEARCH SERVICE. PRODUCTION ECONOMICS RESEARCH BRANCH. Publications issued from December [1953] through April [1954]. no. 1-   Washington, 1954-
Issued bimonthly. A periodical list, which, with the Agricultural Marketing Service's Check list of reports (F4), succeeds the Check

list of printed and processed publications (F8), formerly issued by the U. S. Bureau of Agricultural Economics.

F7 U.S. BUREAU of AGRICULTURAL ECONOMICS. Agricultural economic and statistical publications. Washington, 1952. 83 p.
Complete list, classified by type and subject, of the publications of the Bureau that are available for distribution. Includes descriptions of periodicals and scheduled reports.

F8 U.S. BUREAU of AGRICULTURAL ECONOMICS. Available printed and processed publications issued by the Bureau of Agricultural Economics from July 1, 1940, to February 28, 1941. Washington, 1941. 9 p.
Supplemented monthly by the Agricultural Marketing Service's Check list of reports (F4).

F9 U.S. BUREAU of AGRICULTURAL ECONOMICS. Printed publications issued by the Bureau of Agricultural Economics. Washington, 1943. 17 p.
Lists all printed bulletins of the Bureau of Agricultural Economics from July 1922 to 1943.

F10 U.S. BUREAU of AGRICULTURAL ECONOMICS. LIBRARY. Agricultural economics bibliography. no. 1-97, January 1925-February 1942. Washington, 1925-1942. 97 v.
Complete list of the titles of this series is given on unnumbered pages at the end of no. 97. This series and the Economic library lists (F11) ceased publication when the Bureau of Agricultural Economics Library was made a part of the U. S. Department of Agriculture Library. Economic bibliographies prepared by U. S. Department of Agriculture staff are now usually issued as a Library list (A38) or a Bibliographical bulletin (A35), both of which are numbered series prepared by the U. S. Department of Agriculture Library. Agricultural economic bibliographies have also appeared at times in the various general series of the U. S. Department of Agriculture: a detailed list of these can be found in Important sources of information for work in agricultural economics (F1).

F11 U.S. BUREAU of AGRICULTURAL ECONOMICS. LIBRARY. Economic library list. no. 1-28, March 1939-June 1941. Washington, 1939-1941. 28 v.
Series of bibliographies, each treating a particular phase of agricultural economics, designed to be less comprehensive than those issued in the series, Agricultural economics bibliography (F10). Titles in this series are noted on the unnumbered pages following p. 13 of no. 28. See also notation for F10.

F12 U.S. DEPARTMENT of AGRICULTURE. LIBRARY. Agricultural economics literature. v. 1-16, January 1927-June 1942. Washington, 1927-1942. 16 v.
V. 1-16, no. 3, January 1927-March 1942, prepared by the Bureau of Agricultural Economics Library. Supersedes the Library Supplement to the B. A. E. News; superseded by the Bibliography of Agriculture (A24), issued by the U.S. Department of Agriculture Library. Bibliography of current literature consisting of several sections: 1, Signed reviews; 2, Descriptive notes and abstracts; 3, Bibliographies; 4, Periodical changes; 5, Selected list of recent reviews; 6, U.S. Department of Agriculture publications, economic

in character; 7, State publications; 8, Periodical articles; and 9, Notes (List of current books, pamphlets, etc.). Annual author and subject index.

F13 U.S. PRODUCTION and MARKETING ADMINISTRATION. BUDGET and MANAGEMENT BRANCH. Bibliography of marketing publications issued by the U.S. Department of Agriculture from 1937 through 1946. Washington, 1947. 144 p.
"Prepared as Appendix, Exhibit 3 of report of survey of marketing activities of the Department of Agriculture." Arranged by commodities under marketing phases, such as packaging, transportation, storage and warehousing, processing, etc. In tabular form and gives for each publication the title, date, series, name of author, and issuing agency. Not indexed.

Statistics

The general sources listed below are primarily concerned with international agricultural statistics and statistics of the United States. For a recent list of other official statistical sources the reader is referred to the following publications issued by the Census Library Project of the Library of Congress: Statistical bulletins; an annotated bibliography of the general statistical bulletins of major political subdivisions of the world (F20) and Statistical yearbooks; an annotated bibliography of the general statistical yearbooks of major political subdivisions of the world (F21). Reference should also be made to the section "Important statistical sources for work in agricultural economics" in Important sources of information for work in agricultural economics (F1). A useful guide to governmental statistics is Philip M. Hauser's Government statistics for business use (New York, Wiley, 1956).

F14 FOOD and AGRICULTURE ORGANIZATION of the UNITED NATIONS. Report on the 1950 world census of agriculture. v. 1- Rome, 1955-
V. 1, Census results by countries (loose-leaf, to be added to as various countries complete their census reports). V. 2 and 3 will include a methodological study of the various national censuses and an analysis of the main subjects included in the census. Nearly 100 countries and territories participated in the census.

F15 FOOD and AGRICULTURE ORGANIZATION of the UNITED NATIONS. Yearbook of food and agricultural statistics ... Annuaire de statistiques agricoles et alimentaires. 1947-   Washington and Rome, 1947-
Issued annually. In French and English; table of contents in English, French, and Spanish. Continues the series, International yearbook of agricultural statistics (F17), published by the International Institute of Agriculture. Contains statistics on world agricultural production and livestock numbers. Current statistics appear in the F. A. O.'s Monthly bulletin of agricultural economics and statistics.

F16 INTERNATIONAL INSTITUTE of AGRICULTURE. The first world agricultural census (1930). Rome, 1939. 5 v.
Combines material issued in Bulletins 1-41. Does not include the

USSR and several other countries. Census of each country was carried out by its government and reported according to a standard form issued by the Institute.

F17 INTERNATIONAL yearbook of agricultural statistics. 1910-1945/46. Rome, International Institute of Agriculture, 1912-1947. 27 v.

In French and English. Last issue covers period 1941/42-1945/46. Superseded by F. A. O.'s Yearbook of food and agricultural statistics (F15). A collection of statistical tables showing world agricultural production and numbers of livestock, international trade, and prices of agricultural products.

F18 U.S. BUREAU of the CENSUS. United States census of agriculture: 1950. Washington, 1952. 5 v.

In 5 volumes: v. 1 (34 pts.), Counties and state economic areas; v. 2, General report, statistics by subjects; v. 3, Irrigation of agricultural lands; v. 4, Drainage of agricultural lands; and v. 5, Special reports.

V. 1, comprising state reports, contains statistics on farms, acreage, value, characteristics, livestock and livestock products, crops, and value of farm products. Statistics in v. 2, for the United States and geographic divisions, cover acreage and value of farms, land use, equipment and facilities, livestock, crops, value of farm products, characteristics, farm operators, etc. V. 3 and 4 contain statistics concerning irrigation and drainage. Special reports in v. 5 are: Horticultural specialities, Multiple-unit operations, Ranking agricultural counties, Land utilization; a graphic summary, Farm tenure; a graphic summary, Agriculture 1950; a graphic summary, Irrigation 1950; a graphic summary, Farm-mortgage debt; a cooperative report, Economic class and type of farm; a graphic summary, Farms and farm characteristics by economic subregions.

The first national census of agriculture was taken in 1840, and since then a census of agriculture has been taken in conjunction with the census of population in each decinnial enumeration A mid-decennial agricultural census was initiated January 1, 1925. Thus, since 1920, there has been a federal census of agriculture every fifth year.

For complete information concerning United States census publications from 1790 to date see Catalog of United States census publications issued by the U.S. Bureau of the Census. For a list of censuses of other countries see the U.S. Library of Congress Census Library Project's National censuses and vital statistics in Europe, 1918-1939; an annotated bibliography and Supplement, 1940-1948 (Washington, 1948). See also the same Project's General censuses and vital statistics in the Americas (Washington, 1943).

F19 U.S. DEPARTMENT of AGRICULTURE. Agricultural statistics. 1936- Washington, 1936-

Issued annually. Presents information which before 1936 was published in the statistical section of the Yearbook of agriculture (A81). Tables show acreage, production, prices, supply and distribution of agricultural products, livestock numbers, farm capital and income, foreign trade in agricultural statistics, etc. Includes a table of weights, measures, and conversion factors used in the Department of Agriculture.

Agricultural Economics 165

F20   U.S. LIBRARY of CONGRESS. CENSUS LIBRARY PROJECT.
Statistical bulletins; an annotated bibliography of the general statistical bulletins of major political subdivisions of the world, prep. by Phyllis G. Carter, chief. Washington, 1954. 93 p.

F20a U.S. LIBRARY of CONGRESS. CENSUS LIBRARY PROJECT.
Statistical yearbooks; an annotated bibliography of the general statistical yearbooks of major political subdivisions of the world, prep. by Phyllis G. Carter, chief. Washington, 1953. 123 p.

Legislation

See Section G, Addendum, for additional material.

F21   CODE of federal regulations, containing a codification of documents of general applicability and future effect as of December 31, 1948; with ancillaries and index. Ed. 2, 1949. Ed., pub. by the Division of the Federal Register, the National Archives, pursuant to Section 11 of the Federal register act as amended. Washington, 1949-   50 v. in 45.
Kept up-to-date with pocket supplements and revised volumes. Includes all documents in effect issued by executive branch of the United States government. Contains presidential proclamations, executive orders, orders, regulations, rules, notices, and other instruments issued or promulgated by federal agencies. Current documents of this kind are published in the Federal register.
Titles which are of particular interest to agriculture are: title 6, Agricultural credit; title 7, Agriculture; title 9, Animals and animal products; title 21, Food and drugs; title 36, Parks, forests, and minerals; title 43, Public lands; and title 50, Wildlife.

F22   FOOD and agricultural legislation. v. 1-   1952-   Rome, Food and Agriculture Organization of the United Nations, 1952-
Published in three editions: English, French, and Spanish. An annual compilation of food and agricultural laws and regulations of international importance. A continuation of the Annuaire international de législation agricole (F23), published from 1911 to 1946 by the International Institute of Agriculture.

F23   INTERNATIONAL INSTITUTE of AGRICULTURE. Annuaire international de législation agricole. 1911-1946. Rome, 1912-1948.
An annual summary of important agricultural legislation, arranged by country. Superseded by the F.A.O.'s Food and agricultural legislation (F22).

F24   U.S. DEPARTMENT of AGRICULTURE. OFFICE of INFORMATION.
Abridged list of federal laws applicable to agriculture. (Document 2) 1953. 27 p.
Useful as a brief guide in legislation.

F25   U.S. LAWS, STATUTES, etc. Laws relating to agriculture, comp. by Elmer A. Lewis, Superintendent, Document Room, House of Representatives. Washington, 1950. 576 p.

F26   U.S. LAWS, STATUTES, etc. Revised edition of laws applicable to the United States Department of Agriculture, 1945, containing laws applicable to the Department up to and including the 78th Congress, reorganization plans, executive orders reflecting the origin and organization of agencies of the Department, and current appropriation provisions. Washington, 1945. 2 v.

----- 1945-1950 supplement to laws applicable to the United States Department of Agriculture (1945 edition). Washington, 1952. 898 p.

F27   U.S. OFFICE of EXPERIMENT STATIONS. Federal legislation, rulings, and regulations affecting the state agricultural experiment stations. (U.S.D.A Misc. P. 515, rev.) 1954. 47 p.

RURAL SOCIOLOGY

Rural sociology may seem a somewhat vague and ill-defined field to the uninitiated. Within its purview it includes such subjects as rural social problems, rural-urban differences, health, welfare, religion, recreation, education, and many others which apply to the human rather than the technical phases of farming.

Since, as a subject, rural sociology is sometimes placed with agricultural economics, many of the titles referred to in the preceding section should be kept in mind by the worker in this field. Reference should also be made to general indexes such as the International index to periodicals and the Public affairs information service, both of which index sociological material. The London bibliography of the social sciences (1931-1937; 4 v.) lists over 600,000 items, and supplements are issued irregularly. Social science abstracts lasted only from 1929 to 1933, but gives good coverage for that period. Current sociology (F28), has issued current bibliographies of sociology in alternate issues since 1952.

The rural sociologist must also use the Agricultural index (A23) which indexes under "Sociology, Rural" and related subject headings, articles in agricultural magazines and bulletins issued by state and federal experiment stations and extension services. It also attempts to list books and selected articles from non-agricultural journals which are specifically of interest to the rural sociologist. The monthly and comprehensive Bibliography of agriculture (A24) has a section entitled "Rural social organization," with subsections on human ecology, institutions, leadership, rural organizations and movements, and rural social psychology. Finally, there is the Encyclopedia of social sciences which is a basic tool for all sociologists.

The few works listed below are only the more important of those applying specifically to American rural sociology. A more detailed list of general sociological reference works can by found in Winchell's Guide to reference books (A8) and Kaplan's Research material in the social sciences (Madison, Wis., University of Wisconsin Press, 1939).

Abstracting Journals

F28   CURRENT sociology. 1952-   Paris, UNESCO, 1952-
Issued quarterly. Alternate issues composed of current classified bibliographies of sociology with sections concerning rural sociology.

F29   RURAL sociology; devoted to scientific study of rural life. v. 1- March 1936-   Baton Rouge, La., Rural Sociological Society of America, 1936-
Issued quarterly. Each issue contains book reviews and current

bulletin reviews. The reviews are detailed and by subject specialists. There are no cumulative indexes.

Bibliographies and Indexes

F30   BURCHFIELD, L. Our rural communities, a guidebook to published materials on rural problems. Chicago, Public Administration Service, 1947. 201 p.
Surveys history and development of important areas in rural sociology and provides lists of references for each. Also has a bibliography on "General publications on rural affairs."

F31   LANDIS, B. Y. Guide to the literature on rural life. Ed. 4. New York, Country Life Association, 1939. 15 p.
A selected and briefly annotated bibliography, still useful although somewhat out-of-date.

Directories

F32   U.S. AGRICULTURAL MARKETING SERVICE. Directory of personnel in rural sociology: teachers, research workers, extension workers. Washington, 1955. 31 p.
Arranged by state with an alphabetical index.

Handbooks and Texts

F33   SOROKIN, P. A., ZIMMERMAN, C. G., and GALPIN, C. J., eds. A systematic source book in rural sociology. Minneapolis, University of Minnesota Press, 1930. 3 v.
A work of first importance. "It is intended to be a complete encyclopedia, a reference work, and a substantial systematic treatise in the field." Includes selected readings, many statistics, complete documentation, and bibliographies. World-wide in coverage.

F34   TAYLOR, C. C., and others. Rural life in the United States. New York, Knopf, 1950. 549 p.
A comprehensive survey by specialists. Bibliography, p. 535-549.

AGRICULTURAL EDUCATION

There are very few reference works concerned specifically with agricultural education. However, general compilations such as the Education index (New York, Wilson, 1930-   ) usually cover the subject adequately. This index lists literature on agricultural education in general educational journals as well as that published in such periodicals as the Agricultural education magazine (42). The Encyclopedia of educational research (New York, Macmillan, 1950) has an excellent article on agricultural education plus a good selected bibliography. The Agricultural index (A23) and the Bibliography of agriculture (A24) are also useful.

168 F. Social Sciences

Bibliographies

F35 UNITED NATIONS EDUCATIONAL, SCIENTIFIC, and CULTURAL ORGANIZATION. EDUCATION CLEARING HOUSE. Teaching agriculture: a selected bibliography, prep. by UNESCO and F. A. O. (UNESCO Occasional Papers in Education, no. 15.) Paris, 1952. 52 p.
    An annotated list of world materials considered useful in the teaching of agriculture. Author and subject indexes.

F36 U.S. OFFICE of EDUCATION. Summaries of studies in agricultural education. (Vocational Education B. 180, Agr. Series 17) Washington, 1935.
----- Supplement 1, [1935-1940] Danville, Ill., Interstate, 1943.
----- Supplement 2, [1941-1947] Washington, 1948-
    Issued annually since supplement 3 for 1948. An annotated bibliography of studies in agricultural education with classified subject index. Composed largely of summaries of American graduate theses.

History

F37 TRUE, A. C. A history of agricultural education in the United States, 1785-1925. (U.S.D.A. Misc. P. 36) 1929. 436 p.
    Comprehensive summary of agricultural instruction in schools and colleges, with some reference to research and extension work. Contains a bibliography of 557 references and an index.

F38 TRUE, A. C. A history of agricultural extension work in the United States, 1785-1923. (U.S.D.A. Misc. P. 15) 1923. 220 p.
    Supplements the author's History of agricultural education in the United States (F37). An account of the movement which resulted in the establishment of the system of cooperative extension work in agriculture and home economics. Contains an extensive bibliography and index. Statistics of cooperative extension work, 1914-1923 and the text of the Smith-Lever act are given in the appendix.

Directories

F39 COUNTY agent, vo-ag teacher. 1955 directory. (County agent v. 11, no. 1, Jan. 1955) Philadelphia, Ware Bros., 1955. 100 p.
    Issued annually. Lists for each state, agricultural extension workers, county agricultural agents, county 4-H club agents and vocational agriculture teachers. See also County agents' directory (A66).

F40 INTERNATIONAL INSTITUTE of AGRICULTURE. BUREAU of AGRICULTURAL SCIENCE and PRACTICE. L'enseignement agricole dans le monde. Agricultural education in the world. Rome, 1935-1940. 4 v.
    In 4 parts: pt. 1-2, Europe; pt. 3, North America; pt. 4, Central and South America, Asia, Africa, and Oceania.
    Directory of institutions of agricultural instruction arranged by country, with description of each. Indexed by cities where schools, colleges, etc., are located. In French and English.

F41 U.S. OFFICE of EDUCATION. Directory of schools of agriculture in the Latin American Republics, July 1941. Washington, 1941. 28 p.
List of 182 institutions, giving name, location, and information as to level of instruction offered and specialization, if any. Arranged alphabetically by countries.

Miscellaneous

F42 AGRICULTURAL education magazine. v. 1-   January 1929-   Des Moines, 1929-
Issued monthly. May issues contain annual "Studies in progress in agricultural education," which list graduate theses by students in the field. Book reviews in each issue.

F43 WHAT do studies show? Summaries and interpretations of research in selected areas of agricultural education. Danville, Ill., Interstate, 1952. 69 p.
A series of articles reprinted from the Agricultural education magazine (F42). Bibliographies at end of each article.

SECTION G
ADDENDUM

Reference works noted too late for inclusion in the main body of the text are listed here.

AGRICULTURE - GENERAL

Bibliographies of Bibliographies - General Works

G1   INTERNATIONAL ASSOCIATION of AGRICULTURAL LIBRARIANS and DOCUMENTALISTS. Quarterly bulletin. v. 1-   January 1956-   Harpenden, Herts., Eng., 1956-
"Bibliographical news" section reviews important new agricultural reference works and periodicals. Editor is D. H. Boalch, Librarian, Rothamsted Experimental Station, Harpenden.

G2   OSTVOLD, H. Bibliographic aids for literature research. St. Paul, University of Minnesota. Institute of Agriculture Library, 1955. 20 p.
Designed to instruct agricultural students in bibliographic research. Briefly describes important bibliographic sources for agriculture.

Abstracting Journals

G3   ARGENTINE REPUBLIC. MINISTERIO de AGRICULTURA. DIRECCIÓN de INFORMACIONES. Síntesis bibliográfica, extracto de publicaciones nacionales y extranjeras. Año 1- 1947-   Buenos Aires, 1947-

G4   ČESKÁ zemědělská bibliografie. The Czech agricultural bibliography. Soupis knih a článků. no. 1-   1941-   V. Praze, Brázda, 1941-
Issued by Ústřední zemědělská knihovna ČAZ.

G5   NETHERLANDS. DEPARTEMENT van LANDBOUW, VISSCHERIJ en VOEDSELVOORZIENING. AFDEELING DOCUMENTATIE. Mededelingen. 's-Gravenhage, 1945-
Issued weekly.

G6   POLJOPRIVREDNA bibliografija FNRJ. The agricultural bibliography of Yougoslavia. no. 1-   1945-   Beograd, 1947-
Issued annually by Ministarstvo Poljoprivrede FNRJ.

Bibliographies and Indexes

G7 BUENOS AIRES. UNIVERSIDAD NACIONAL. FACULDAD de
    AGRONOMÍA y VETERINARIA. BIBLIOTECA. Compilacíon
    de las bibliografías existentes en la Biblioteca de la Faculdad
    de Agronomia y Veterinaria. Buenos Aires, 1951. 30 p.
G8 FOOD and AGRICULTURE ORGANIZATION of the UNITED NATIONS.
    Bibliography on land and water utilization and use in Europe, by
    C. H. Edelman and B. E. P. Eeuwens. Rome, 1955. 347 p.
    Covers agricultural geography, climate, soils, vegetation, hydrology
    and hydrography, land reclamation, drainage, irrigation, soil
    erosion, and soil conservation. Each section arranged geographically.
G9 GREAT BRITAIN. MINISTRY of AGRICULTURE and FISHERIES.
    LIBRARY. A selected and classified list of books relating to
    agriculture, horticulture, etc., in the Library of the Ministry
    of Agriculture and Fisheries. Ed. 3. London, 1954. 89 p.

Encyclopedias

G10 ENCICLOPEDIA agraria italiana. Pubblicata sotto gli auspici della
    Federazione Italiana dei Consorzi Agrari. Roma,
    Editoriale degli Agricoltori, 1952-
G11 VEENMAN'S agrarische Winkler Prins; encyclopedie vor landbouw,
    tuinbouw en bosbouw, onder redactie van G. Minderhoud [et al].
    Wageningen, Veenman, 1954-

Dictionaries

G12 TODD, E. F. H. Diccionario técnico-agrícola; inglés-espanol. English-
    Spanish technical-agricultural dictionary. [Mexico] 1953. 99 p.

PLANT SCIENCES

HORTICULTURE AND AGRONOMY

Bibliographies and Indexes

G13 SCHOLL, A. W. and CARTWRIGHT, D. R. Bibliography of synthetic
    plant-growth substances. Huntington, W. Va., A. W. Scholl (669
    S. Terrace, Huntington 5, W. Va.), 1955. 532 p.
    Based entirely on Chemical abstracts (D8).

Dictionaries and Encyclopedias

G14 PAREY'S illustriertes gartenbaulexikon. Hrsg. von Richard Maatsch
    unter mitwirkung zahlreicher mitarbeiter. Ed. 5. Berlin,
    Parey, 1955-
    When completed will be one of the most comprehensive horticultural dictionaries in any language.

FORESTRY AND FOREST PRODUCTS

Dictionaries

G15   AMELINCKX, F. Lexicon dendrologicum. (Latin-Dutch-French-English-American-German) The Hague, Nijhoff, 1955. 528 p.

ANIMAL SCIENCES

VETERINARY MEDICINE

Classification Systems

G16   BERNARD, C. C. A classification for medical and veterinary libraries. Ed. 2. London, Lewis, 1955. 279 p.
A revision and expansion of the Bernard classification scheme used in the Libraries of the London School of Hygiene and Tropical Medicine and several other British medical libraries.

ECONOMIC ENTOMOLOGY

Reviews

G17   ANNUAL review of entomology. 1-   1956-   Stanford, Calif., Annual Reviews, 1956-
Critical reviews of important developments in entomology. Extensive bibliographies.

COMMERCIAL FISHING AND FISHERIES

Abstracting Journals

G18   U. S. FISH and WILDLIFE SERVICE. Sport fishery abstracts. v. 1-   July 1955-   Washington, 1955-
Issued quarterly. Abstracts current literature in sport fishery research and management and papers on limnology, ecology, natural history, and other fields when concerned with sport fish and fishing.

Bibliographies and Indexes

G19   U. S. FISH and WILDLIFE SERVICE. Fishery publication index, 1920-1954. (C. 36) 1955. 254 p.
Subject, author, and series index of U. S. Bureau of Fisheries publications, 1920 to 1940, and fishery publications of the U. S. Fish and Wildlife Service from 1940 to 1954.

G20   U. S. FISH and WILDLIFE SERVICE. Thirty-five year index of the fishery technological service publications of interest to fishery technologists, by M. E. Stansby. Seattle, 1953. Pts. I-III.
Lists by author and subject publications of the U. S. Bureau of

Fisheries and the U. S. Fish and Wildlife Service from 1918 to 1953. Brings (C484) down through 1953.

PHYSICAL SCIENCES

AGRICULTURAL ENGINEERING

Directories

G21 FOOD and AGRICULTURE ORGANIZATION of the UNITED NATIONS. Directory of wheel and track-type tractors produced throughout the world. Rome, 1955. 371 p.

METEOROLOGY

Bibliographies and Indexes

G22 SCHNEIDER, M. Agrarmeteorologische bibliographie, 1950. (Bibliographien des Deutschen Wetterdienstes 1) 1955. 141 p.

FOOD AND NUTRITION

General Works, Reviews, and Bibliographies of Bibliographies

G23 PETERSON, M. S. A guide to the literature on food research and technology. In U. S. Quartermaster Food and Container Inst. Activities rpt. 6(1954):16-26.
Particularly useful for selected list of food science journals.
G24 SHANK, R. E. and MRAK, E. M. A bookshelf on foods and nutrition. In Amer. J. of Pub. Health 45(1955):419-428.
A critical analysis of recent and important sources of information for the food sciences.

Bibliographies and Indexes

G25 DIAZ, H. S. de. Nutritional aspects and other uses of food yeasts; annotated bibliography, 1917-1949. (Puerto Rico Agr. Expt. Sta. B. 92) 1951. 241 p.
G26 STEPHANY, C. D. and VON LOESEKE, H. W. Selected references on yeast. Washington, War Food Administration, Office of Marketing Services, 1945. 353 p.

Dictionaries

G27 FISCHER, W. Fachwörterbuch für brauerei und mälzerei. Ed. 2. Teil I: Deutsch-Englisch. Teil II: Englisch-Deutsch. Nuremberg, 1955. 2 v.

Periodical Lists

G28 "List of periodicals scrutinized." In Food Science Abstracts 26(1954):IV-VIII.
See also G23 for selected list of food science journals.

SOCIAL SCIENCES

AGRICULTURAL ECONOMICS

Legislation

G29 U.S. DEPARTMENT of AGRICULTURE. Compilation of statutes relating to soil conservation, marketing quotas and allotments, crop insurance, sugar payments and quotas, price support, Commodity Credit Corporation, and related statutes as of January 1, 1955. (Agr. Handb. 79) 1955. 193 p.

G30 U.S. LAWS, STATUTES, etc. Farm relief and Agricultural adjustment acts, comp. by Elmer A. Lewis. Washington, 1954. 473 p.

G31 U.S. LAWS, STATUTES, etc. Federal farm loan act with amendments, and farm mortgage and farm credit acts, comp. by Elmer A. Lewis. Washington, 1950. 271 p.

INDEX

Abbreviations of periodical titles, A108, A122
Aberdeen-Angus cattle, C128-C131
Abetz, K. and Köstler, J., eds. Forstliche rundschau, B201
Abstracts of bacteriology, B158
Abstracts of canning technology, E6
Abstracts of recent published material on soil and water conservation, D24
Acarology, C329
Acridological abstracts, C292
Adressbuch der Deutschen nahrungs- und genussmittelbetriebe, E94
Advances in agronomy, B82
Advances in food research, E1
Agassiz, L. Bibliographia zoologiae et geologiae, C11
Agrar-Bibliographie, A9
Agrarmeteorologische bibliographie, 1950, M. Schneider, G22
Agricultural and horticultural engineering abstracts, National Institute of Agricultural Engineering (Gt. Brit.), D55
Agricultural chemistry
  abstract journals, D6-D11
  bibliographies and indexes, D12-D14
  dictionaries, D16-D17
  encyclopedias, D15-D17
  handbooks, manuals, textbooks, D18-D21
  history, D20

  literature guides, D1-D5
Agricultural economics
  bibliographies and indexes, F2-F13
  guides to literature, F1
Agricultural education
  bibliographies and indexes, F35-F36
  directories, F39-F41
  history, F37-F38
  theses: bibliographies, F36, F42
Agricultural education magazine, F42
Agricultural engineering
  abstract journals, D54-D55
  bibliographies and indexes, D56-D63
  dictionaries, D64
  directories, D65-D75, G21
  institutions: directory, D68
Agricultural engineering (periodical), p. 136
Agricultural engineers' yearbook, 1955, D65
Agricultural experiment stations
  abstract journals, A21-A22
  check list of publications...on the subject of plant pathology, 1876-1920, B178
  directories, A70-A71, A76, A78-A79
  history, A93
  laws and legislation, F27
  list of bulletins...from their establishment to 1942, A43
  list of publications on soils, D29
  poultry publications: bibliography, C182-C185

[175]

Agricultural extension work
   directories, F39
   history, F38
Agricultural geography see Agriculture. Atlases
Agricultural index, A23
Agricultural library directory, A69
Agricultural library notes, U. S. Department of Agriculture. Library, A36
Agricultural machinery see Agricultural engineering
Agricultural pest control see Pest control
Agricultural research in U. S. Dept. of Agriculture, A95
Agricultural society directories, A66, A73-A74, A123-A126
Agricultural statistics, U. S. Department of Agriculture, F19
Agriculture
   abstract journals, A9-A22, G3-G6
   atlases, A100-A107
   bibliography, G8
   bibliographies and indexes, A1-A8, A23-A46, G1-G2, G7-G9
   biography, A31-A32, A83-A91, A97
   dictionaries, A53-A61, G12
   directories, A62-A79
   economic aspects: bibliography, A92
   encyclopedias, A47-A52, G10-G11
   experimentation: history, A93
   geography see Agriculture. Atlases
   handbooks, manuals, textbooks, A80-A82, E138
   history, A84, A86-A90, A92-A94, A98
   bibliography, A86, A90, A92
   laws and legislation, F21-F27, C461-C463, G29-G31
   literature guides, G2
   periodicals, A108-A122
   statistics, A127, A129, F14-F20
   study and teaching see Agricultural education
   tropical: bibliography, A27

Agronomy
   abstract journals, B83-B86
   annual reviews, B82
   bibliographies and indexes, B98, B100, D28
   handbooks, manuals, textbooks, B125-B126
   See also Millet, Sorghum, Barley, Oats, Rice, Rye, Tobacco, Wheat, Cotton
Agronomy current literature, U. S. Bureau of Plant Industry, Library, B98
AHD series 4, U. S. Bureau of Animal Industry, p. 77
——— series 6, U. S. Bureau of Animal Industry, p. 77
Ahern, G. P. and Newton, H. K. A bibliography on woods of the world, B203
Ainsworth, G. C. The plant diseases of Great Britain, B165
Ainsworth, G. C. and Bisby, G. R. A dictionary of the fungi, B189
Allen, F. P. A check list of periodical literature and publications of learned societies of interest to zoologists in the University of Michigan libraries, C57
Allen, J. M. and Barnes, H. V. A bibliography of plant pathology in the tropics and in Latin America, B167
Allen, L. F. American cattle, their history, breeding, and management, C106. History of the Short-horn cattle, C142
Alley, H. The national beekeepers' directory, C425
Allyn, R. A dictionary of fishes, C500
Alphandéry, E. Encyclopédie apicole, C412
Amelinckx, F. Lexicon dendrologicum, G15
Amelung, H., ed. Die pflanzenstoffe botanisch-systematisch bearbeitet, B66
American Aberdeen-Angus Breeders' Association. Origin of the Aberdeen-Angus and its

American Aberdeen-Angus Breeders' Association - (continued) development in Great Britain and America, C128
American Association for the Advancement of Science. Zoological names, C29
American Association of Economic Entomologists. Common names of insects approved by the American Association of Economic Entomologists, C344. Membership list, C347
American beet sugar companies, 1953-1954, E95
American brewer register. The brewery buyer's guide, E96
American Can Company. The canned food reference manual, E135. Library abstracts, E7
American Chemical Society. Division of Chemical Literature. Literature resources for chemical process industries, D1, E2. Searching the chemical literature, D2
American Dietetic Association. Journal, E8
American Entomological Society, Philadelphia. Catalogue of works in the library..., C303. Transactions, v. 1, C303
American Fisheries Society. A list of common and scientific names of the better known fishes of the United States and Canada, C501. List of members, C514
American Fisheries Society history, C535
American Home Economics Association. Food and Nutrition Division. Terminology Committee. Handbook of food preparation, E136
American Institute of Biological Sciences. Advisory Committee on Hydrobiology. Directory of hydrobiological laboratories and personnel in North America, C515
American Joint Committee on Horticultural Nomenclature. Standardized plant names, B102
American Meat Institute, Chicago. Department of Public Relations. Reference book of the meat industry, E97
American Medical Association. Council on Foods and Nutrition. Handbook of nutrition, E137
American men of science, A83
American Meteorological Society. Meteorological abstracts and bibliography, D76
American Paper and Pulp Association. Dictionary of paper, including pulps, boards, paper properties, and related papermaking terms, B226
American Poultry Association. The American standard of perfection, C190
American Society of Refrigerating Engineers. Refrigeration abstracts, E9
American standard of perfection, American Poultry Association, C190
American Veterinary Medical Association. Directory, C264. —— Committee on Revision of Veterinary Anatomical Nomenclature. Nomina antomica veterinaria, C249
Amerine, M. A. and Wheeler, L. B. A check list of books and pamphlets on grapes and wine and related subjects, 1938-1948, E28
Ammal, E. K. J. and Darlington, D. C. Chromosome atlas of cultivated plants, B155
Ammonium in soils, bibliography, D34
Andrew, J., rev. Agricultural dictionary, English-Polish and Polish-English, A59
Anell, B. Contribution to the history of fishing in the southern seas, C534
Angling see Fishing and fisheries

Animal-breeding
  abstract journals, C64-C65
  bibliography, C23, C81
  history, C115
Animal breeding abstracts, Commonwealth Bureau of Animal Breeding and Genetics, C64
Animal diseases see Livestock, Diseases; Veterinary medicine; Poultry, Diseases; Cattle, Diseases; Horses, Diseases
Animal geography see Zoology, Economic. Atlases and geography and Animal husbandry. Geography
Animal husbandry
  abstract journals, C64-C65
  bibliographies, C66-C83
  dictionaries, C89-C94
  directories, C95-C98
  encyclopedias, C84-C88
  geography, C157-C163
  periodical lists, p. 83
  research institutions in Europe, C95
Animal industry
  abstract journals, C65
  bibliography, C66, C82-C83
  directories, C95-C98
  history, C106-C127
  prose and poetry, C165
Animal nutrition see Livestock. Feeding
Annales agronomique, Series A, L'Institut National de la Recherche Agronomique, A14
Annales mycologici editi in notitiam scientiae mycologicae universalis, B159
Annuaire des docteurs en médecine vétérinaire de Belgique C265
Annuaire fructidor ... annuaire international des fruits, légumes, primeurs, dérivés et industries annexes, 1952, B111
Annuaire national de l'agriculture, 1945-1946, A62
Annuaire vétérinaire, C266
Annual meat packers' guide, 1955, E98
Annual review of biochemistry, D6

Annual review of entomology, G17
Annual review of plant physiology, B4
Annuario vinicolo d'italia, E99
Apicultural abstracts, C400
Apiculture see Bee culture
Appel, O., ed. Handbuch der pflanzenkrankheiten, B186
Applegate, V. C., Macy, P. T., and Harris, V. E. Selected bibliography on the applications of electricity in fishery science, C477
Apstein, C. and Wasikowski, K. Periodica zoologica, C58
Arber, Mrs. A. Herbals, their origin and evolution, B68
Arboretum directories, B51, B117
Archiv für bienenkunde, C401
Arctic Institute of North America. Arctic bibliography, C17
Arctic regions, bibliography, C17
Argentine Republic. Ministerio de Agricultura. Dirección de Informaciones. Síntesis bibliográfica, extracto de publicaciones nacionales y extranjeras, G3
―――― Ministerio de Agricultura y Ganderia de la Nacion. Departamanto de Bibliotecas. Boletín bibliográfico, A25
Arid zone research, directory of institutions engaged in, A73
Arnold Arboretum publication no. 3, B28
Aro, P., and others. Suomalais, ruotsalais, saksalais, englantilainen metsäsanakirja ... , B227
Artschwager, E. F. Dictionary of biological equivalents, German-English, C30
Artschwager, E. F. and Smiley, E. M. Dictionary of botanical equivalents, B33
Ashby, H., and others. German-English botanical terminology, B34
Aslin, M. S. Library catalogue of printed books and pamphlets on agriculture published between 1471 and 1840, A33

L'Association des Ingenieurs
  Forestiers de la Province
  de Quebec. Vocabulaire
  forestier, B228
Association of American Pesticide
  Control Officials. Pesticide
  official publication and con-
  densed data on pesticide
  chemicals, C446
Association of British Insecticide
  Manufacturers. Directory,
  C443
Association of Land-Grant Col-
  leges and Universities.
  Engineering Division. Engi-
  neering division record, D56
Association of Official Agricultural
  Chemists. Official methods
  of analysis, D18
Associations see Agricultural
  societies and subheading
  Societies under specific sub-
  jects
Atanasoff, D. Virus diseases of
  plants, B166
Atlas international zootechnique,
  International Institute of
  Agriculture, C159
Atlas of American agriculture, U.
  S. Department of Agricul-
  ture, A105
Atlas of climatic types in the
  United States, 1900-1939, C.
  W. Thornthwaite, A104
Atlas of the world's resources, W.
  Van Royen, A107. v. III:
  Forest and fishery resources,
  B274
Atwood, A. C. and Blake, S. F.
  Geographical guide to flora
  of the world, B53
Australian pastoral directory; list
  of stockowners, p. 77
Avery, T. B. and Payne, L. F.
  International poultry
  guide for flock selection,
  C193
Avian egg, The, A. L. and A. J.
  Romanoff, C209
Aviation entomology, bibliography,
  C437
Aviculture see Birds, Poultry

Babcock, E. B. and Clausen, R. E.
  Genetics in relation to agriculture, B154
Backer, C. A. Verklarend woorden-
  boek der Wetenschappelijke
  namen van de in Nederland en
  Nederlandsch-indie in het wild
  groeiende en in tuinen en
  parken geweekte varens en
  hoogere planten, B35
Bacon, L. B., and others. Agricultural geography of Europe
  and the Near East, A100
Bacteriology
  abstract journals, B158, B164
  handbooks, manuals, textbooks,
    B185
Bading, H. Wörterbuch der landwirtschaft, A54
Bague, J. Glosario de biologia
  animal (español-inglés), C92.
  Glossary of animal biology
  (English-Spanish), C93
Bailey, L. H. Cyclopedia of American agriculture, A47. Manual
  of cultivated plants, most commonly grown in the continental
  United States and Canada,
  B103. Standard cyclopedia of
  horticulture, B104
Bailey, L. H. and E. Z. Hortus
  second, B105
Bailey, S. F. and Smith, L. M.
  Handbook of agricultural pest
  control, C447
Baillière's encyclopaedia of scientific agriculture, H. Hunter,
  A48
Baillon, H. Dictionnaire de botanique, B36. Histoire des
  plantes, B54
Baker, E. T. The home veterinarian's handbook, C243
Baker, O. E. Atlas of American
  agriculture, A105. A graphic
  summary of American agriculture based largely on the
  census, C157. A graphic summary of farm animals and
  animal products (based largely
  on the census of 1930 and
  1935), C158. Geography of the
  world's agriculture, A106

Bakers' digest, E10
Baking industry
  abstract journals, E10
  directories, E100
  literature guide, E2
Baking industry (periodical). Bakers' buying directory issue, 1954, E100
Balfour, I. P., rev. History of botany, 1530-1860, B72
Banks, N. Directions for collecting and preserving insects, C393. A list of works on North American entomology, C309, C310. See also U. S. Bur. of Entomology. Bibliography of the more important contributions to American economic entomology
Baranowski, Z. The international horseman's dictionary, C89
Baranski, A. Die vorgeschichtliche zeit im lichte der hausthiercultur, C107
Barley bibliography, B135
Barnard, C. C. A classification for medical and veterinary libraries, G16
Barnes, H. V. and Allen, J. M. A bibliography of plant pathology in the tropics and in Latin America, B167
Bartholomew, J. G. Atlas of zoogeography, C51
Bawden, F. C. Plant viruses and virus diseases, B182
Bay, J. C. Bibliographies of botany, a contribution toward a bibliotheca bibliographica, B1
BDI-Inf series 4, U. S. Bureau of Dairy Industry, p. 77
——— series 6, U. S. Bureau of Dairy Industry, p. 77
Beaufort, L. F. de. Zoogeography of the land and inland waters, C52
Bedevian, A. K. Illustrated polyglottic dictionary of plant names in Latin, Arabic, Armenian, English, French, German, Italian, and Turkish languages, B37

Bee culture
  abstract journals, C400-C401
  bibliography, C399, C401-C411
  dictionaries, C416-C424
  directories, C425-C427
  encyclopedias, C412-C415
  history, C428-C430
  library catalogs, C403, C405-C411
  periodicals, C399, C403, C411, C429, C431-C433
  societies, C415
Beebe, C. W. A monograph of the pheasants, C194
Beef bibliography, E49
Beef cattle industry, tropical, bibliography, C66
Beekeepers see Bee culture
Beet sugar industry directory, E95
Beilstein, F. Handbuch der organischen chemie, D12
Beltramini de Casati, F. Vocabolario apistico italiano e dizionario d'apicoltura, C416
Bennett, H. The chemical formulary, D19
Bentham, G. and Hooker, J. D. Genera plantarum ad exemplaria imprimis in herbrariis Kewensibus servata definita, B55
Bercaw, L. O. The poultry industry, C171
Bergey's manual of determinative bacteriology, Society of American Bacteriologists, B185
Berichte über die gesamte biologie, A10
Berichte über die gesamte physiologie und experimentelle pharmakologie, A10
Berichte über die wissenschaftliche biologie, A10
Bericht über die wissenschaftlichen leistungen im gebiete der entomologie, C294
Berlin. Biologische Zentralanstalt für Land- und Forstwirtschaft in Berlin-Dahlen. Bibliographie der pflanzenschutzliteratur, B168
Bessey, E. A. Morphology and taxonomy of fungi, B190

Besterman, T. Index bibliographicus, A4. A world bibliography of bibliographies, A1
Better farming methods, p. 77
Bezemer, T. J. Dictionary of terms relating to agriculture, horticulture, forestry, cattle breeding, dairy industry, and apiculture, in English, French, German, and Dutch, A55
Bibliographia forestalis, B202
Bibliographia genetica, B133
Bibliographia oceanographica, C471. Publications périodiques consultées pour la préparation de la bibliographie, C552. Vocabulorum elenchus prior lexico totius scientiae oceanographicae redigendo, C502
Bibliographia zoologica, C6
Bibliographic index, A2
Bibliographie der pflanzenschutzliteratur, Berlin. Biologische Zentralanstalt für Land- und Forstwirtschaft in Berlin-Dahlen, B168
Bibliographie entomologique, C. Nodier, C301
Bibliographies of bibliographies, A1-A8
Bibliographies on pure and applied botany and related subjects, B2
Bibliography of cerealiana, C. H. Briggs, E30
Bibliography of poultry diseases, C238
Bibliography of soil science, fertilizers, and general agronomy, 1931-1950, Commonwealth Bureau of Soil Science, D28
Bibliography of systematic mycology, Commonwealth Mycological Institute, B169
Bibliography on woods of the world, exclusive of the temperate region of North America and with emphasis on tropical woods, B203

Bibliotheca entomologica, H. A. Hagen, C299
Bibliotheca zoologica. I, C12; II, C13
Bibliographic index, A2
Bidwell, P. W. and Falconer, J. I. History of agriculture in the northern United States, 1620-1860, A84
Biedermann's zentralblatt für agrikulturchemie und rationellen landwirtschaftsbetrieb, D7
Bien, C. Tropical beef cattle industry in the Western Hemisphere, C66
Bierer, B. W. American veterinary history, C272. History of animal plagues of North America, C273. A short history of veterinary medicine in America, C274
Biochemistry see Agricultural chemistry
Biography see Agriculture. Biography
Biological abstracts, A11
Biological laboratory directories, A68
Biological sciences serial publications, U. S. Library of Congress. Reference Department. Science Division, A119
Bibliography of agriculture, A24 classification scheme, A135
Birds
    bibliographies, C18, C24, C27, C166
    nomenclature, C44
Bisby, G. R. and Ainsworth, G. C. A dictionary of the fungi, B189
Bitting, K. Gastronomic bibliography, E29
Black's veterinary dictionary, W. C. Miller and G. P. West, C257
Blake, S. F. Guide to popular floras of the United States and Alaska, B52
Blake, S. F. and Atwood, A. C. Geographical guide to flora of the world, B53
Blanchard, R. A. É. Glossaire allemand-francais des terms

Blanchard, R.A.E. (continued)
   d'anatomie et de zoologie, C31
Blanck, F. C. Handbook of food and agriculture, E138. Nutritional data, E154
Blanco, G. and Varela, A. Publicaciones periodicas agricolas en America Latina, A121
Blüher, P. M. Meisterwerk der speisen und getränke, E73
Bly, F. Verklarende vakwoordenlijst van de zeevisscherij, C503
Boalch, D. H., ed. Quarterly bulletin, International Association of Agricultural Librarians and Documentalists, G1
Bodenheimer, F. S. Materialien zur geschichte der entomologie bis Linné, C376
Bois, D. Les plantes alimentaires chez tous les peuples et à travers les ages, B61
Boletín bibliográfico agricola, Spain. Ministerio de Agricultura, A20
Borgman, W., Brödermann, E. V., and Gisevius, P. Illustriertes landwirtschafts lexikon, A50
Boston, E. J. Jersey cattle, C140
Botanical abstracts, B5
Botanical gardens
   bibliography, B11
   directories, B49, B51, B117
   history, B11
Botanical museums
   bibliography, B11
   directories, B49
   history, B11
Botanical review, B6
Botanische jahrbücher für systematik pflanzengeschichte und pflanzengeographie, B7
Botanisches centralblatt, B8
Botany
   abstract journals, B5, B7-B8
   annual reviews, B4, B6, B9
   bibliographies and indexes, B1-B32

   biography, B69, B73
   classification, B3, B14-B16, B18, B26, B32, B41
   dictionaries, B33-B48
   economic, B61-B67, B102
   history, B68-B73
   international congresses, B49
   nomenclature, B75, B102
   physiology, B4
   pictorial works, B19
   societies, B11, B49
Botany current literature, U. S. Bureau of Plant Industry. Library, B30
Boulton, E. H. B. A dictionary of wood, B229
Bower, W. T. History of the American Fisheries Society, C535
Bowes, A. De P. and Church, C. F. Food values of portions commonly used, E150
Branconnier, R. and Glandard, J. Nouveau Larousse agricole, A49
Bradley, A. V. Tables of food values, E151
Bradley, M. A. and Hunt, M. G. Index to the publications of the United States Department of Agriculture, 1901-1925, A40
Bradley bibliography, The, A. Rehder, B28
Breed, R. S., Murray, E. G. D., and Hitchens, A. P. Bergey's manual of determinative bacteriology, B185
Breeders' gazette, p. 77
Breeding see Plant-breeding and Animal-breeding
Brewers' digest, E11
Brewers' digest annual buyers' guide and directory, E101
Brewing industry
   abstract journals, E11, E17, E27
   bibliographies and indexes, E56
   dictionaries, E79, G27
   directories, E96, E101, E115
Bridges' food and beverage analyses, M. R. Mattice, E143
Briggs, C. H. A bibliography of cerealiana, E30
Briggs, H. M. Modern breeds of livestock, C99

Briquet, R. Pequene dicionário inglês-português de têrmos empregados em anatomia, ezoognósia, fisiologia, zootecnia e tecnologia dos produtos animais, C90

British Museum (Nat. Hist.) Library. Catalogue of the books, manuscripts, maps, and drawings in the British Museum (Natural History), B13 and C7

British poultry standards, C191

British Standards Institution. British standard specifications, B230. Recommended common names for pest control products, C441

Britten, J. and Boulger, G. S. A biographical index of deceased British and Irish botanists, B69

Britten, J. and Holland, R. Dictionary of English plant names, B38

Britton, N. L. and Shafer, J. A. North American trees, being descriptions and illustrations of the trees growing independently of cultivation in North America, north of Mexico, and the West Indies, B258

Brödermann, E. U., Gisevius, P., and Borgman, W. Illustriertes landwirtschafts lexikon, A50

Brooklyn Entomological Society. Explanation of terms used in entomology, C331

Brooks, R. M. and Olmo, H. P. Register of new fruit and nut varieties, 1920-1950, B112

Brosch, A. Die literatur über das schwein von 1538 bis zur gegenwart, C67. Das schrifttum über das rind, C68. Verzeichnis der bienenliteratur des in- und auslandes 1890-1918, C402

Brown, A. W. A. Insect control by chemicals, C448

Browne, C. A. A source book of agricultural chemistry, D20

Brückmann, F. E. Bibliotheca animalis oder verzeichnis der meisten schriften so von thieren und deren theilen handeln, C223

Brues, C. T. Classification of insects, C394. Insects and human welfare, C364

Brunet, R. Dictionnaire d'oenologie et de viticulture, E74

Bruttini, A. Dictionnaire de sylviculture en cinq langues, B231

Buchholz, E. Kleines russischdeutsches forstwörterbuch, B232

Buenos Aires. Universidad Nacional. Faculdad de Agronomía y Veterinaria. Biblioteca. Compilacíon de las bibliografías existentes en la Biblioteca de la Faculdad de Agronomia y Veterinaria, G7

Bulletin, Public Affairs Information Service, p. 77, 166

Bulletin of zoological nomenclature, International Committee on Zoological Nomenclature, p. 71

Bundesanstalt für Forst- und Holzwirtschaft, Reinbek. Bibliographie des ausländischen forst- und holzwirtschaftlicher schrifttums, B204

Burchfield, L. Our rural communities, F30

Burkill, I. H. A dictionary of the economic products of the Malay Peninsula, B62

Business directories, A67

Butler, Sir E. J. Plant pathology, p. 50

Buttress, F. A. Agricultural periodicals of the British Isles, 1681-1900, and their location, A109

Cable, M. Introduction to ornithological nomenclature, C44

Cagny, P. Dizionario veterinario, C250

Cagny, P. and Cobert, H. J. Dictionnaire vétérinaire, C251
Campbell, D. M. and Merillat, L. A. Veterinary military history of the United States, C282
Canada. Department of Agriculture. Index to entomological publications of the Department of Agriculture, 1884-1936, C314. Guide to the chemicals used in crop protection, C449
Canada food directory, 1946-1947, E102
Canadian dairy and ice cream journal. Dairy industry buyers' directory, E103
Canadian fisheries annual, 1955, C516
Candolle, A. and C. de. Monographiae phanerogamarum, B56
Candolle, A. P., A., and C. de. Prodromus systematis naturalis regni vegetabilis, B57
Candy industry catalog and formula book, 1954/55, E104
Canned food reference manual, American Can Company, E135
Canners' directory, 1954, National Canners' Association, E116
Canning and preserving
 abstract journals, E6-E7
 handbooks, manuals, textbooks, E135, E139
 literature guide, E2
Canning trade directories, E116, E128, E130
Canning trade almanac, E139
Cannon, M. A. and Handy, R. B., comps. List by titles of publications of the United States Department of Agriculture from 1840 to June 1901, A41
Card, L. E. A check list of poultry bulletins, C182, C183
Caribou bibliography, C21
Carlander, K. D. Handbook of freshwater fishery biology, C524

Carnegie Endowment for International Peace. Division of International Law. Monograph series no. 7, C558
Carnegie Institution of Washington. Publication no. 358, A84. Publication no. 430, A88
Carnegie Library of Pittsburgh. Bibliography of physical properties and bearing value of soils, D26
Carpenter, J. R. An ecological glossary, B39, C32
Carpenter, M. M. Bibliography of biographies of entomologists, C386, C387
Carter, P. G. Statistical bulletins, F20. Statistical yearbooks, F20a
Cartwright, D. R. and Scholl, A. W. Bibliography of synthetic plant-growth substances, G13
Catalog list of American bee journals, C431
Catalogue of scientific papers, 1800-1900, Royal Society of London, C16
Caton, D. Selected bibliography - range resources and management, B87
Cattle
 bibliography, C66, C68-C71, C81, C83
 breeds, C106-C145
 diseases, C273, C276
 encyclopedias, C84-C88
 genetics, C81
 geography, C159
 history, C106-C145
 See also Livestock, Animal-breeding, etc.
Cattle plague see Rinderpest
Cereals and cereal products bibliography, E30
Česká zemědělská bibliografie, G4
Chamberlin, W. J. Entomological nomenclature and literature, C290
Chapman, D. H. A farm dictionary, A56
Chapman, W. M., comp. Translations of fisheries literature from foreign languages into English, C497

Chase, A. Manual of the grasses of the United States, B74
Chatfield, C. Food composition tables for international use, E152. Food composition tables, minerals, and vitamins for international use, E153
Chemical abstracts, D8
Chemical analysis, official methods, D18
Chemical formulary, H. Bennett, D19
Chemisches zentralblatt, D9
Chemistry see Agricultural chemistry
Chicago. University. Institute of Meat Packing. Books and pamphlets on the meat packing industry, E31
Chickens see Poultry
Chilean Nitrate Educational Bureau, Inc. Bibliography of the literature on the minor elements and their relation to plant and animal nutrition, D27
Chittenden, F. J. Dictionary of gardening, B106
Christ, J. L. Algemeines theoretisch-praktisches wörterbuch über die bienen und die bienenzucht, C417
Chromosomes, B155
Chronica botanica. World list international congresses, societies, etc., B49
Church, C. F. and Bowes, A. De P. Food values of portions commonly used, E150
Claassen, E. S. An annotated check list of the more important entomological periodicals, C389
Classification of books
  agricultural, A134, G16
  bibliography of agriculture, A135
  U. S. Dept. of Agriculture. Library, A136
  veterinary, G16
Clausen, R. E. and Babcock, E. B. Genetics in relation to agriculture, B154
Clawson, M. The western range livestock industry, C108
Clay, E. W. List of available publications of the United States Department of Agriculture, A26
Clemen, R. A. American livestock and meat industry, C109
Climate see Meteorology
Climate and man, U. S. Department of Agriculture, D85
Clute, W. N. The useful plants of the world, B63
Cobacho, J. G. Diccionario de términos técnicos de veterinaria, C252
Cobert, H. J. and Cagny, P. Dictionnaire vétérinaire, C251
Code of federal regulations, F21
Coffee trade directories, E126
Colcord, M. Check list of publications on entomology issued by the United States Department of Agriculture through 1927, C315
Collingwood, G. H. Knowing your trees, B259
Colloidal materials in soils, abstract journals, D23
Coman, E. T. Sources of business information, p. 160
Comité International du Bois. Yearbook of world timber trade, 1938, B278
Commercial fertilizer yearbook, 1954, D48
Commercial fisheries abstracts, U. S. Fish and Wildlife Serv., C467
Commerical fishing see Fishing and fisheries
Commonwealth Agricultural Bureaux. Executive Council. Gazetteer of agricultural and forestry research stations in the British Commonwealth, 1952, A63. List of research workers, 1956, A64
Commonwealth Bureau of Agricultural Parasitology (Helminthology). Helminthological abstracts, C3

Commonwealth Bureau of Animal Breeding and Genetics. Animal breeding abstracts, C64. Technical communication no. 8, C85

Commonwealth Bureau of Animal Health. Index veterinarius, C221. Veterinary bulletin, C218

Commonwealth Bureau of Animal Nutrition. Nutrition abstracts and reviews, E13

Commonwealth Bureau of Dairy Science. Dairy science abstracts, E12

Commonwealth Bureau of Horticulture and Plantation Crops. Horticultural abstracts, B83

Commonwealth Bureau of Pastures and Field Crops. Field crop abstracts, B84. Herbage abstracts, B85

Commonwealth Bureau of Plant Breeding and Genetics. Plant breeding abstracts, B129

Commonwealth Bureau of Soil Science. Bibliography of soil science, fertilizers, and general agronomy, 1931-1950, D28. Soils and fertilizers, D22

Commonwealth Forestry Bureau. The chief research centres for forestry and wood utilization in continental Europe, excluding most of the Balkan Peninsula and Eastern Europe, B243. Forestry abstracts, B200

Commonwealth Institute of Entomology. Distribution maps of insect pests. Series A, C388. The review of applied entomology. Serial publications examined, C390. Review of applied entomology. Series A: Agricultural, C293. Review of applied entomology Series B: Medical and veterinary, C219

Commonwealth Mycological Institute. Bibliography of systematic mycology, B169. Distribution maps of plant diseases, B181. Index of fungi, B191. Mimeographed publications no. 7-8, B169. The review of applied mycology, B160

Comstock, J. H. Introduction to entomology, C365

Confédération Européene de l'Agriculture. Publications de la CEA. Fasc. 5, A65

Confédération Européene de l'Agriculture. Vade-mecum des principales organisations internationales tenant compte spécialement de leurs rapports avec l'agriculture et des organisations agricoles internationales et institutions internationales apparentées à l'agriculture, A65

Congreso Panamericano de Medicina Veterinaria. I, Lima, 1951. Contribución a la bibliografía veterinaria americana, C232

Conley, A. R. History of research dealing with problems confronting the dairy industry which have been conducted by the agricultural experiment stations, 1881-1947, E32

Connor, L. G. A brief history of the sheep industry in the United States, C146

Conservation directories, D49

Conservation of wild life see Wildlife. Conservation

Conservation yearbook, C47, D49

Conversion factors, A128, A131-A133

Cook, M. T. Host index of virus diseases of plants, B170. Index of the vectors of virus diseases of plants, B171

Cook, M. T. and Otero, J. I. A bibliography of mycology and phytopathology of Central and South America, Mexico, and the West Indies, B176. Partial bibliography of virus diseases of plants, B177

Cook-books, bibliography, E36, E38,

Cook-books (continued)
E50-E53, E70
Cookery see Food
Coon, E. L. List of serials currently received in the Library of the United States Department of Agriculture, A117
Copenhagen. K. Veterinaer- og Landbohøjskole. Bibliotek. Katalog over den Kongelige veterinaer- of landbohøjskoles bibliotek, indtil udgangen af 1894, C225
Corkhill, T. A glossary of wood, B233
Cotton literature, U. S. Department of Agriculture. Library B101
Cottonseed oil industry, directory, E112
Coues, E. American ornithological bibliography, C18
County agent, vo-ag teacher, F39
County agents' directory, A66
County agricultural agents directories, A66, F39
Crane, E. E. Dictionary of beekeeping terms, C418
Crane, E. J. and Patterson, A. M. Guide to the literature of chemistry, D3
Croker, R. S. Glossary of Japanese fishery terms, C504
Cultivated plants, see Plants, cultivated
Cummings, O. E. Important sources of information for work in agricultural economics with special emphasis on California, F1
Current list of medical literature, U. S. Armed Forces Medical Library, p. 90, 126
Current literature in agricultural engineering, U. S. Bureau of Agricultural Chemistry and Engineering. Library, D61
Current sociology, F28
Cuthbertson, S. A preliminary bibliography on the American fur trade, C19

Cyclopedia of American agriculture, L. H. Bailey, A47

Dahl, J. O. Dictionary of 1001 menu terms, E75
Dairy cattle see Cattle
Dairy industries catalog of equipment, supplies, and services used by dairy products manufacturers, E105
Dairy Industries Society, International. French and English glossary of dairy and related terms, E76. Spanish and English glossary of dairy and related terms, E76a
Dairy industry
  directories, E103, E105, E132
  literature guide, E2
Dairy science
  abstract journals, E12, E20
  bibliographies and indexes, C83, E32, E149
  dictionaries, E76-E77, E87
  periodicals, E149
Dairy science abstracts, Commonwealth Bureau of Dairy Science, E12
Dairying bibliography, C83
Dalla Torre, K. W. von and Harms, H. Genera siphonogamarum, B58
Dalton, B. Sources of engineering information, p. 136
Dansk veterinaerhistorisk aarbog, C275
Darlington, D. C. and Ammal, E. K. J. Chromosome atlas of cultivated plants, B155
Daubert, B. F. Bibliography on the deterioration of fat in foods, E68
Davis, E. G. Rural electrification in the United States, D57
Davis, H. S. Culture and diseases of game fishes, C525
Davis, J. G. A dictionary of dairying, E77
Davis, J. J. The entomologists' joke book, C395
Davis, J. R. and Duncan, H. S. History of the Poland China

Davis, J.R. and Duncan, H.S. (Cont'd) breed of swine, C151
Davis, M. V. Guide to American business directories, A67
Dayton, W. A. Glossary of botanical terms commonly used in range research, B40. United States tree books, B205
Dean, B. A bibliography of fishes, C472
Deen, J. L., Benson, A. B., and Dannfelt, J. J. A Swedish-English vocabulary for foresters, B234
Dehydration of food see Food, Dried
De la Torre-Bueno, J. R. A glossary of entomology, C332
Dennis, R. W. G. Plant diseases in orchard, nursery, and garden crops, B183
De Ong, E. R. Insect, fungus, and weed control, C450
Dethier, V. G. Chemical insect attractants and repellents, C451
Deutsche landwirtschaftliche rundschau, A12
Deutsches Entomologisches Museum. Katalog der bibliothek, C304
Deutschlands weinbauorte und weinbergslagen, E106
Devries, L. French-English science dictionary for students in agricultural, biological, and physical sciences, A57. German-English science dictionary for students in chemistry, physics, biology, agriculture, and related sciences, A58
Diaz, H. S. de. Nutritional aspects and other uses of food yeasts; annotated bibliography, 1917-1949, G25
Dickson, J. G. Diseases of field crops, p. 50
Dictionaries, bibliography, A53. See also subheading Dictionaries under specific subjects
Dieckerhoff, W. Geschichte der rinderpest und ihrer literature, C239
Dietetics
 abstract journals, E8
 handbooks, manuals, textbooks, E142
Dihydrostreptomycin, C242
Directory of biological laboratories, A68
Directory of international scientific organizations, United Nations Educational, Scientific and Cultural Organization, A74
Directory of organization and field activities of the Department of Agriculture: 1955, U. S. Department of Agriculture, A76
Diseases of field crops, J. G. Dickson, p. 50
Distillation literature, index and abstracts, A. and E. Rose, D11
Doe, J. and Marshall, M. L. Handbook of medical library practice, C216
Dog diseases, C260
Dowling, E. C. and Gill, T. The forestry directory, B246
Drake, C. J. Bibliography of the state plant pest laws, quarantines, regulations and administrative rulings of the United States of America, C396
Drug plants
 dictionary, B46
 nomenclature, B102
Ducks, C198
Duerst, J. U. Grundlagen der rinderzucht, C69
Duerst, R. J. Die rinder von Babylonien, Assyrien, und Ägypten und ihr zusammenhang mit den rindern der alten welt, C110
Dumont, P. A. National wildlife refuge bibliography, C20
Duncan, H. S. and Davis, J. R. History of the Poland China breed of swine, C151
Dutilly, A. A. A bibliography of reindeer, caribou, and musk-ox, C21

Dyer, Mrs. A. I. (Robertson).
Guide to the literature of
home and family life, E33
Dzieduszycki, A. Agricultural
dictionary, English-Polish
and Polish-English, A59

Eastman Kodak Company,
Rochester, N. Y. Distillation
Products Industries. Annotated bibliography of vitamin E, E34
Ecological crop geography, K. H.
W. Klages, A103
Ecological glossary, An. J. R.
Carpenter, B39
Economic botany see Botany,
economic
Economic entomology
abstract journals, C292-C293
annual reviews, G17
bibliographies, C294-C330
bibliographies of bibliographies,
C1, C290-C291
biography, C385-C387
dictionaries, C331-C343
directories, C346-C362
diseases of insects, C328
geography, C388
government publications,
C294-C299
handbooks, manuals, textbooks,
C363-C375
history, C298, C376-C384
laws, C396
library catalogs, C283-C288
museums, collections, C397
nomenclature, C10, C290, C344,
C345
periodicals, C389-C392
research institution directory,
C361
society directories, C354, C361
Economic ornithology see Poultry
husbandry
Economic zoology see Zoology,
Economic
Edelman, C. H. and Eeuwens, B.
E. P. Bibliography on land
and water utilization and use
in Europe, G8
Edible plants see Plants, edible

Education index, p. 167
Edwards, E. E. Bibliography of the
history of agriculture in the
United States, A86
Edwards, E. E. and Rasmussen, W.
D. A bibliography on the agriculture of the American
Indians, A85
Eeuwens, B. E. P. and Edelman, C.
H. Bibliography on land and
water utilization and use in
Europe, G8
Egg industry, directory, E133
Eggs see Poultry industry
Eiselt, J. N. Geschichte, systematik und literatur der insectenkunde, von den ältesten
zeiten bis auf die gegenwart,
C298 and C377
Ekman, S. P. Zoogeography of the
sea, C551
Electricity in fishery science bibliography, C477
Elenchus librorum de apium cultura, A. de Keller, C404
Ellenberger-Schütz jahresbericht,
p. 91
Elsevier's encyclopaedia of organic
chemistry, D15
Empire Forestry Association. The
Empire forestry handbook,
1952, B244
Enciclopedia agraria italiana, G10
Encyclopaedia of poultry, C186
Encyclopedia Britannica, p. 3
Encyclopedia of educational research, p. 167
Encyclopedia of the social sciences,
p. 160, 166
Encyclopédie biologique, B61
Engelmann, W. Bibliotheca
historico-naturalis, C14
Engineering see Agricultural engineering
Engineering index, D54
Engler, A. Syllabus der pflanzenfamilien; mit besonderer
berücksichtigung der nutzpflanzen nebst einer übersicht
über die florengebiete der
erde, B14
Engler, A. and Prantl, K. Die
natürlichen pflanzenfamilien

Engler, A. and Prantl, K. (cont'd.) nebst ihren gattungen und wichtigeren arten insbesondere den nutzpflanzen, B15
English-Russian agricultural dictionary, N. F. Geminova, T. C. Krasnosselskaja, and B. N. Ussowsky, A60
Ensminger, M. E. The stockman's handbook, C100
Entoma, C444
Entomological news, p. 100, 110
Entomological Society of America. List of members, C348
Entomological technique (ET-series) publications, C317
Entomologische literaturblätter, C295
Entomologischer Verein, Stettin. Katalog der bibliothek, C305
Entomology see Economic entomology
Entomology current literature, U. S. Bureau of Entomology and Plant Quarantine. Library, C297
Erdman, F. S. Bibliography of literature on frozen foods, E35
Erosion see Soil erosion
Essig, E. O. A history of entomology, C378
Ewing, W. R. Poultry nutrition, C206
Experiment station record, U. S. Office of Experiment Stations, A22
Experiment stations see Agricultural experiment stations
Extension work see Agricultural extension work

Faber, F. L. Bibliography on poultry and egg marketing, C172
Faes, H. Lexique viti-vinicole international, E78
Falconer, J. I. and Bidwell, P. W. History of agriculture in the Northern United States, 1620-1860, A84

Family life bibliography, E33
Farlow, W. G. Bibliographical Index of North American fungi, B192
Farm animals see Livestock
Farm chemicals handbook, D50
Farm implement news' 1954 buyer's guide issue with implement repair information, D66
Farm implements see Agricultural engineering
Farm relief and Agricultural adjustment acts, U. S. Laws, Statutes, etc., G30
Farmers' organizations see Agricultural society directories
Farmer's yearbook encyclopaedia and directory of pedigree breeders, p. 77
Farming see Agriculture
Farming and mechanised agriculture, A80
Fat in foods, deterioration of: bibliography, E68
Fearing, D. B. Check list of books on angling, fish, fisheries, fish-culture, etc., C488
Featherly, H. I. Taxonomic terminology of the higher plants, B41
Federal farm loan act with amendments, and Farm mortgage and Farm credit acts, U. S. Laws, Statutes, etc., G31
Feeds and feeding, C102-C103
Feeds and feeding, F. B. Morrison, C102
Feige, E. Die haustierzonen der alten welt, C111
Feldkamp, C. L. and Pennington, C. E. List of the publications on soils issued by the state agricultural experiment stations of the United States through 1926, D29
Fenton, F. A. Field crop insects, C366
Fernald, H. T. Applied entomology, C367
Fern dictionary, B48
Fertilization of fishponds, C481-C482
Fertilizers
  abstract journals, D22, D25

Index   191

Fertilizers (continued)
  directories, D48, D50
Field crop abstracts, Commonwealth Bureau of Pastures and Field Crops, B84
Field crops see Agronomy
Field Museum of Natural History. Zool. ser. v. 25, pts. 1-3, C166
Finch, V. C. and Baker, O. E. Geography of the world's agriculture, A106
Fischer, G. W. The smut fungi, B193
Fischer, V. E. List of publications on apiculture contained in the U. S. Department of Agriculture Library and in part those contained in the Library of Congress, C403
Fischer, W., and others. Fachwörterbuch für brauerie und mälzerie, E79, G27
Fisher, H. J., and others, eds. Official methods of analysis, Association of Official Agricultural Chemists, D18
Fisher, R. A. and Yates, F. Statistical tables for biological, agricultural, and medical research, A127
Fisheries year-book and directory, C517
Fisherman's encyclopedia, C505
Fishes
  bibliography, C472, C476
  nomenclature, C501, C508, C510
Fishing and fisheries
  abstract journals, C467-C470, G18
  bibliographies, C464-C499, G19-G20
  dictionaries, C500-C513
  directories, C514-C523
  doctoral dissertations, C483
  electricity in, C477
  encyclopedias, C500-C513
  handbooks, manuals, textbooks, C524-C533
  history, C534-C550
  industrial wastes affecting, C472

  laws and regulations, C544, C557-C558
  literature guides, C465
  motion pictures, C498-C499
  periodicals, C552-C553
  research and educational institutions, C520, C523
  statistics, C554-C556
  translations of foreign literature, C496-C497
Fishing gazette annual review ... , C518
Fishing tackle digest, C519
Flavor of foods, bibliographies and indexes, E61, E69
Fleece see Wool bibliography
Fleming, G. Animal plagues, C276
Flint, W. P. and Metcalf, C. L. Destructive and useful insects, C370
Flock books, C79
Flood control bibliography, D35
Floras, B52-B60, B103, B106
Floriculture see Horticulture
Florists see Horticulture. Directories
Flour and feed trade
  directories, E117-E118
  literature guide, E2
Flowering plants dictionary, B48
Fontaine, A. Nouveau dictionnaire vétérinaire, médicine, chirurgie, thérapeutique, législation sanitaire et sciences qui s'y rapportent, C253
Fontaine, H. Das deutsche heeresveterinärwesen, C277
Food
  abstract journals, E6-E27
  bibliographies and indexes, E28-E72
  chemistry and analysis, E140-E141, E143, E148, E150-E158
  dictionaries and encyclopedias, E73-E93
  handbooks, manuals, textbooks, E136, E138
  laws and legislation see Agriculture. Laws and legislation
  literature guides, E1-E5, G23-G24
  periodicals, G23, G28

Food (continued)
  research, E145
  statistics see Agricultural
    statistics
  value and composition tables,
    E150-E158
  See also Nutrition
Food, Dried, bibliography, E64
Food, Frozen
  bibliographies, E35, E71
  directories, E109, E120
Food and agricultural legislation,
    F22
Food and Agricultural Organization of the United Nations. Bibliography of forestry and forest products, B206. Bibliography on land and water utilization and use in Europe, G8. Directory of forestry schools, B245. Directory of wheel and track type tractors produced throughout the world, G21. FAO forestry and forest product studies 7, B275. Fisheries technology literature, C464. Food and agricultural legislation, F22. Forestry abstracts coverage list, periodicals, and serials, B275. List of plant breeders in Canada and the United States of America, B151. Handbook for World fisheries abstracts, C469. Home economics information exchange, E3. Index of agricultural research institutions in Europe concerned with animal production and their principal lines of investigation, C95. Report on the 1950 world census of agriculture, F14. Timber statistics for the years ... , B279. World list of plant breeders, B152. Yearbook of forest product statistics, B280. Yearbook of food and agricultural statistics, F15
―― Nutrition Division. FAO nutritional studies no. 3, 11, E152-E153. Food composition tables for international use, E152. Food composition tables, minerals and vitamins for international use, E153
―― Secretariat. Fisheries Div. Commodity standards for fisheries products, C557. F.A.O. fisheries study no. 2, C557. Fisheries and related subjects; list of films, C498. Fishery research and educational institutions in North and South America, C520
Food habits, E145
Food industries catalogs for food processors, E107
Food industries manual, E80
Food industry
  directories, E94-E134
  literature guide, E2, E65, G23, G24
Food packer, annual buyers' guide and reference issue, October 15, 1954, E108
Food plants see Plants, edible
Food science abstracts, Great Britain. Department of Scientific and Industrial Research, E16
Food technology, E14
Forage and pasture crops, W. A. Wheeler, B126
Forage crops see Agronomy
Forbes, R. D. and Meyer, A. B. eds. Forestry handbook, B270
Ford Robertson, F. C. The Oxford system of decimal classification for forestry, B283
Foreign plant diseases ... a manual of economic plant diseases which are new to or not widely distributed in the United States, J. A. Stevenson, B187
Forest products
  bibliography and indexes, B206, B212-B214, B216-B221
  directories, B248-B257
  statistics, B278-B282
Forestry
  abstract journals, B199-B201
  atlases, B274
  bibliographies and indexes, B202-B225
  decimal classification system, B283

Forestry (continued)
dictionaries, B226-B242, G15
directories, B243
handbooks, manuals, texts, B267-B273
periodicals, B275-B277
statistics, B278-B282
subject headings, B284-B285
Forestry abstracts, Commonwealth Forestry Bureau, B200
Forestry current literature, U. S. Forest Service. Library, B222
Forestry handbook, Society of American Foresters, B270
Forestry research directories, B243-B247
Forestry schools see Forestry research directories
Formulas, D19
Forschungsdienst, Der. A13
Forstliche rundschau, B201
Fortschritte der botanik, B9
France. Office de la Recherche Scientifique et Technique Outre-Mer. Bulletin signaletique d'entomologie médicale et vétérinaire, C236
Franck, W. J. and Bruijning. General seed bibliography, B88
Fraser, H. M. Beekeeping in antiquity, C428
Frauendorfer, S. Von. Internationale bibliographie der agrarökonomischen schrifttums, F2. Nouvelles sources bibliographiques agricoles, A5. Système de classification des sciences agricoles, A134
Frear, D. E. H. Agricultural chemistry, a reference text, D21. A catalogue of insecticides and fungicides, C452. Chemistry of the pesticides, C453
French-English science dictionary for students in agricultural, biological, and physical sciences, L. Devries, A57

Friedmann, H. The birds of North and Middle America. Part X: Family Cracidae ... , C195
Froehner, R. Kulturgeschichte der tierheilkunde, C278
Frozen food see Food, Frozen
Frozen food factbook and directory, 1954, E109
Fruit annual, 1952-1953: year-book and directory of the world's fruit trade, B113
Fruit trade directories, B111-B113, B118
Fruit varieties, B107, B112
Frykholm L. Förteckning över bitidskrifter, C432. Förteckning över löpande utländska seriepublikationer vid svenska bibliotek inom lantbrukets, veterinärmedicinens, skogshushållningens, jaktens och fiskets ämnesområden, A110. List of specimen copies of foreign agricultural serials excluding bee and dairy periodicals in the Library of the Royal Agricultural College of Sweden, A111. Översikt över lantbruksforskningens och dess hjälpvetenskapers viktigare bibliografiska hjälpmedel, A3. Översikt över viktigare bilitterature, bibliografiska hjälpmedel och forsknings-institutioner, C399
Fungi
abstract journals, B159-B162
bibliographies and indexes, B169, B173-B176, B192-B195, B220
classification, B190-B191, B195-B196
dictionaries, B189, B197
handbooks, manuals, textbooks, B198
Fungus control
handbooks, manuals, textbooks, C450, C452
laws and regulations, C446, C461-C463
Fur-bearing animals
abstract journals, C4-C5
bibliographies, C23, C25, C28

Fur-bearing animals (continued)
dictionaries, C40
Fur trade bibliography, C19

Gaines, S. H. Bibliography on soil erosion and soil water conservation ... , D30. Glossary of terms used in soil conservation, D44
Game and game-birds see Wildlife, Birds
García-Piquera, C. Glosario de terminologia forestal, B235
Gardening see Horticulture
Garrett, T. F. The encyclopaedia of practical cookery, E81
Gartenbauwissenschaft, B86
Gastronomic bibliography, K. Bitting, E29
Gastronomy
  bibliographies and indexes, E29, E36, E38, E52-E53, E58, E70
  encyclopedias, E89
Gaul, A. T. A glossary of terms and phrases used in the study of the social insects, C333
Gazetteer of agricultural and forestry research stations in the British Commonwealth 1952, Commonwealth Agricultural Bureaux. Executive Council, A63
Geiger, R. The climate near the ground, D83
Geisendorf, A. L. Vade-mecum des principales organisations internationales tenant compte spécialement de leurs rapports avec l'agriculture et des organisations agricoles internationales et institutions internationales apparentées à l'agriculture, A65
Geminova, N. F., Krasnosselskaja, T. C., and Ussowsky, B. N. English-Russian agricultural dictionary, A60
General Foods. Research and Development Department. Library. Current abstracts, E15

Genetics see Plant-breeding and Animal-breeding
Geographical guide to flora of the world, S. F. Blake and A. C. Atwood, B53
Geography of the world's agriculture, U. S. Department of Agriculture, A106
Georgia. Engineering Experiment Station. Special report 23, E71
German-English botanical terminology, H. Ashby, and others, B34
German-English dictionary for foresters, O. Raber, B239
German-English science dictionary for students in chemistry, physics, biology, agriculture, and related sciences, L. Devries, A58
Germany. Statistisches Reichsamt. Deutscher landwirtschaftsatlas, A101
Gerth van Wijk, H. L. Dictionary of plant names, B42
Ghidini, G. M. Glossario di entomologia, C334
Giannini Foundation of Agricultural Economics. Library. Important sources of information for work in agricultural economics with special emphasis on California, F1
Gill, T. and Dowling, E. C. The forestry directory, B246
Gisevius, P., Borgman, W., and Brödermann, E. H. Illustriertes landwirtschafts lexikon, A50
Gistel, J. Die jetzt lebenden entomologen, kerffreunde und kerfsammler Europa's und deren übrigen continente, C349. Lexikon der entomologischen welt, der carcinologischen und arachnologischen, C350. Die naturforscher diess- und jenseits der oceane, C351
Glandard, J. and Braconnier, R. Nouveau Larousse agricole, A49

Glossary of special terms used in the soils yearbook, D45

Goat bibliography, C71

Gooch, D. W. Bibliography on the marketing of livestock, meat, and meat products, C70

Goode, G. B. The fisheries and fishery industries of the United States, C536

Goodspeed, C. E. Angling in America, C537

Gourley, J. E. An annotated list of bibliographies of cookery books, E51. Eating round the world, foreign recipe books and magazine articles in English, E36. Regional American cookery, 1884-1934, E52

Government statistics for business use, P. M. Hauser, p. 163

Gow, R. M. The Jersey; an outline of her history during two centuries - 1734-1935, C141

Graf, D. W. Agricultural engineering, D58. Irrigation, D59

Graham, S. A. Forest entomology, C368

Grain trade buyers' guide, 1954, B114

Grain trade directory, B114

Gram, E. and Weber, A. Plant diseases in orchard, nursery, and garden crops, B183

Grapes see Viticulture and Wines and wine-making

Gras, N. S. B. A history of agriculture in Europe and America, A87

Grasses, B74, B80. See also Agronomy

Grasshopper control directory, C360

Graves, L. G. and Taber, C. W. A dictionary of food and nutrition, E82

Gray, L. C. and Thompson, E. K. History of agriculture in the southern United States to 1860, A88

Gray herbarium card index, B16

Great Britain. Colonial Office. Fishery publication 5, C482. Pesticide abstracts and news summary, C434

Great Britain. Colonial Products Laboratory. Consultative Committee on Insecticide Materials. Bibliography of insecticide materials of vegetable origin, C436.

—— Department of Scientific and Industrial Research. Food science abstracts, E16

—— Meteorological Office. The meteorological glossary, D81

—— Ministry of Agriculture and Fisheries. Library. A selected and classified list of books relating to agriculture, horticulture, etc., in the Library of the Ministry of Agriculture and Fisheries, G9

Green, J. R. A history of botany, 1860-1900, B70

Gries, C. G. Foreign weights and measures with factors for conversion to United States units, A132

Grocery trade directories, E119, E125

Grünwoldt, F. Die dokumentation in der forstwirtschaft, B207. Internationales adressbuch für forstwirtschaft, holzwirtschaft, B247. Répertoire international des périodiques forestiers, sylviculture, économie du bois, protection de la nature et chasse d'après leur état au 1er janvier 1940, B276

Guayule bibliography, D38

Guernsey cattle, C132-C133

Guide to reference books, C. M. Winchell, A8

Guilletmot, A. Förteckning över periodiska publikationer rörande mejerihanteringen E149

Hackh, I. W. D. Hackh's chemical dictionary (American and British usage), D16

Hadders, M., rev. Die pflanzenstoffe botanisch-systematisch bearbeitet, B66

Hagen, H. A. Bibliotheca entomologica, C299

Hainsworth, R. G. Agricultural geography of Europe and the Near East, A100. A graphic summary of world agriculture, A102

Hall, D. G. How to get further information on insects, C346

Haller, H. L. Common names of insecticides for use in the Journal of Economic Entomology, C442

Hamlin, H. M., ed. American farmers' and rural organizations, A124

Hammond, J. Progress in the physiology of farm animals, C164

Hampton, J. F. Modern angling bibliography, C489

Handwörterbuch für bienenfreunde, C419

Handy, R. B. and Cannon, M. A. comps. List by titles of publications of the United States Department of Agriculture from 1840 to June 1901, A41

Hanlin, H. H., Weller, W. W., and Fagg, C. J., comps. The stowage red book, A130

Hannan, A. M. Influence of weather on crops: 1900-1930, D77

Hanover (City). Tierärztliche Hochschule. Bibliothek. Katalog, C226

Hansen, J. J. P. Den Danske dyrlaegeforening, 1849-1949, C279

Hardy, A. C. Seafood ships, C526

Harding, T. S. Two blades of grass, A89

Harms, H. and Dalla Torre, K. W. von. Genera siphonogamarum, B58

Harper's directory and manual, 1952: the standard work of reference for the wine and spirit trade, E110

Harris, P. L. and Kujawski, W. Annotated bibliography of vitamin E, E34

Harvard University. Arnold Arboretum. Library. Catalogue, B17

—— Museum of Comparative Zoology. Bulletin v. 108, C394

Harvey, R. B. An annotated bibliography of the low temperature relations of plants, B89

Hatcheries see Poultry industry

Hatchery and feed, p. 88

Hatchery tribune and feed retailer, p. 88

Hauser, P. M. Government statistics for business use, p. 163

Hawaiian Sugar Planters' Association. Experiment Station. Division of Entomology. Report of work. Bulletin no. 8, C227

Hawes, I. L. Bibliography on aviation and economic entomology, C437

Hawks, E. B. A bibliography of poultry, C173. Cattle, sheep, and goat production in the range country, C71. List of agricultural periodicals of the United States and Canada published during the century July 1810 to July 1910, A115

Hawks, E. B. and Trolinger, C. List of the publications on soils issued by the U. S. Department of Agriculture, 1844-1926, D31

Hay-fever plants, B81

Hay trade directory, B119

Hayes, H. K., and others. Methods of plant breeding, B156

Hazelton, J. M. History and hand book of Hereford cattle and Hereford bull index, C134

Heald, F. D. Manual of plant diseases, p. 49, 50

Heckel, M. and Zander, R. Wörterbuch der gärtnerischen fachausdrücke in vier sprachen, B110

Hedrick, U. P. Cyclopedia of hardy

Hedrick, U. P. (continued) fruits, B107. A history of horticulture in America to 1860, B127. Sturtevant's notes on edible plants, B65. The vegetables of New York, B108

Hedwigia, B161

Heinz (H. J.) Company. Nutritional data, E154

Hekmeyer, F. C. Korte geschiedenis der runderpest, benevens eene opgave van al de over deze ziekte handelende geschriften, die van de vroegste tijden tot op heden zijn uitgekommn, C240

Helfer, J. W. Terminologia entomologica, C335

Helminthological abstracts, Commonwealth Bureau of Agricultural Parasitology (Helminthology), C3

Helminthology abstract journals, C3

Henderson, I. F. Dictionary of scientific terms, C33

Henderson, J., rev. Bibliography on soil conservation, D42

Hennefrund, H. E. Bibliography on the poultry industry in countries other than Canada and the United States, C174

Henshaw, Samuel see U. S. Bur. of Entomology. Bibliography of the more important contributions to American economic entomology

Henze, J. K. G. Entwurf eines verzeichnisses veteranischer schriften und einzelner abhandlungen, C224

Herbage abstracts, Commonwealth Bureau of Pastures and Field Crops, B85

Herbage crops see Agronomy

Herbals, their origin and evolution, Mrs. A. Arber, B68

Herbaria directory, B50

Herd books, C79

Hereford cattle, C134-C138

Hering, M. Lepidopterologisches wörterbuch, C336

Herms, W. B. Medical entomology, C369

Hess, R. W. and Record, S. J. Timbers of the New World, B269

Hesse, R. Ecological animal geography, C53

Hewston, E. M., and others. Vitamin and mineral content of certain foods affected by home preparation, E155

Hickey, J. J. and Wagner, F. H. A check list of technical game bulletins published by the state game departments, C22

Hill, C. L. The Guernsey breed, C132

Hirsch-Schweigger, E. Zoologisches wörterbuch, C34

Historia agriculturae, A90

Hitchcock, A. S. Manual of the grasses of the United States, B74

Hoffman, A. Entomologen-adressbuch, C352

Hogner, D. C. Farm animals and working and sporting breeds of the United States and Canada, C101

Hohman, E. P. The American whaleman, C538

Holden, W. and Wycoff, E. Bibliographical contributions from the Lloyd Library, B24

Holland, J. H. Overseas plant products, B64

Holland, R. and Britten, J. Dictionary of English plant names, B38

Holman, H. P., and others. Index of publications of the Bureau of chemistry and soils, originally the Bureau of chemistry and Bureau of soils, v. 1, D14

Holmes, F. O. Handbook of phytopathogenic viruses, B184

Holstein-Friesian cattle, C139

Home and family life bibliography, E33

Home economics
   abstract journals, E21
   bibliographies and indexes, E3, E4, E33, E62

Home economics (continued)
   education, E66
   information sources, E3
   theses, E63
Hook, I. E. Irrigation engineering, p. 136
Hooker, J. D. and Bentham, G. Genera plantarum ad exemplaria imprimis in herbrariis Kewensibus servata definita, B55
Hopper, E. G. List of periodicals currently received in the Library of the United States Department of Agriculture, June 1, 1936, A116
Hormone bibliography, G13
Horn, W. and Kahle, I. Über entomologische sammlungen, ein beitrag zur geschichte der entomo-museologie, C397
Horn, W. and Schenkling, S. Index litteraturae entomologicae. Ser. 1: Die welt-literatur über die gesamte entomologie bis inklusive 1863, C300
Horsemanship dictionary, C89
Horses
   diseases, C247
   history, C153-C156
Horticultural abstracts, Commonwealth Bureau of Horticulture and Plantation Crops, B83
Horticultural guide, 1954, B115
Horticultural trade directory, B116
Horticulture
   abstract journals, B83, B86, D55
   bibliographies and indexes, B92, B97
   dictionaries, B105-B106, B110, G14
   directories, B111, B113, B115-B118, B122-B124
   encyclopedias, B104, B109
   history, B127
   nomenclature, B102
   periodicals, B128
Hydrobiological laboratories directory, C515
Hortus second, L. H. and E. Z. Bailey, B105

Host index of the fungi of North America, A. B. Seymour, B196
Host index of virus diseases of plants, M. T. Cook, B170
Houck, U. G. The Bureau of Animal Industry of the United States Department of Agriculture, C112
Hough, R. B. Handbook of the trees of the northern states and Canada east of the Rocky Mountains, B260
Howard, A. L. A manual of timbers of the world, B267
Howard, L. O. A history of applied entomology (somewhat anecdotal), C379
Huet, M. Petit glossaire limnologique, C506
Hunter, H. Baillière's encyclopaedia of scientific agriculture, A48
Hunt, M. G. and Bradley, M. A. Index to the publications of the United States Department of Agriculture, 1901-1925, A40
Hutt, F. B. Genetics of the fowl, C207
Hybridization see Plant-breeding
Hydrobiology bibliographies, C471, C473, C475
Hydrography dictionary, C507

Ice cream abstract journals, E20
Ichthyology see Fishing and fisheries
Illustriertes landwirtschafts lexikon, A50
Imker ABC, Das: nachschlagewerk für alle gebiete der bienenzucht, R. Jacoby, C413
Imkers wortschatz, K. Puhlemann, C424
Imperial Bureau of Animal Breeding and Genetics. Bibliography on fur breeding, C23
Imperial Bureau of Animal Genetics. Bibliography on the biology of the fleece, C72
Imperial Bureau of Plant Breeding and Genetics. Bibliography on cold resistance in plants, B134. Bibliography on insect

Imperial Bureau of Plant Breeding and Genetics (continued)
   pest resistance in plants (with a supplement on resistance to nematodes), C326
Imperial Bureau of Plant Genetics. Barley breeding bibliography, B135. Bibliography on the breeding and genetics of the millets and sorghums, B136. Breeding resistant varieties, 1930-1933, B137. Breeding varieties resistant to disease, B138. The experimental production of haploids and polyploids, B139. Interspecific and intergeneric hybridization in relation to plant-breeding, B140. Oat breeding bibliography, B141. Rice breeding bibliography, B142. Rye breeding bibliography, B143. Tobacco breeding bibliography, B145. Wheat breeding bibliography, B144
Implement and tractor catalog file, 1954, D67
Index bibliographicus, T. Besterman, A4
Index-catalogue of medical and veterinary zoology, U. S. Bureau of Animal Industry, C237
Index-catalogue of the library of the Surgeon General's Office, United States Army ... authors and subjects, U. S. Army Medical Library, C9
Index kewensis, B18
Index litteraturae entomologicae. Ser. 1: Die weltliteratur über die gesamte entomologie bis inklusive 1863, W. Horn and S. Schenkling, C300
Index londinensis to illustrations of flowering plants, ferns, and fern allies, B19
Index of fungi, Commonwealth Mycological Institute, B191
Index Society. Publication VIII, B10
Index to American Botanical literature, Torrey Botanical Club, B29

Index to American mycological literature, Mycologia, B175
Index to publications by the International Council for the Exploration of the Sea, 1899-1938, International Council for the Study of the Sea, C479
Index to the literature of American economic entomology, C311
Index veterinarius, Commonwealth Bureau of Animal Health, C221
Indians of North America, agriculture, A85
Industrial arts index, D60
Industrial research laboratories of the United States, National Research Council, A72
Industrial wastes and fishing bibliography, C478
Ingram, W. M. Publications on industrial wastes relating to fish and oysters; a selected bibliography, C478
Innis, H. A. The cod fisheries, C539
Insect control see Insecticides and Pest control
Insect pest resistance in plants
   bibliography, C326
   textbook, C371
Insecticide abstracts and news summary. Great Britain. Agricultural Research Council. Interdepartmental Insecticide Committees, C434
Insecticides
   abstract journals, C434-C435
   bibliographies, C436, C438
   directories, C443-C445, C457
   handbooks, manuals, textbooks, C446-C460
   laws and regulations, C446, C461-C463
   terminology, C442
Insects
   classification, C374-C375, C394
   collecting and preserving, C393, C398
   collections, C397
   diseases of, C328
   See also Economic entomology
L'Institut National de la Recherche Agronomique. Series A, B, C, E, A14

Institute for Government Research Service. Monograph of the U. S. government no. 50, C463; no. 59, C383; no. 60, C382
Institute of Brewing, London. Journal of the Institute of brewing, E17
Institute of Food Technologists. Directory of members, 1956, E111
Institute of Paper Chemistry. Library. Bulletin, B199
Inter-American Institute of Agricultural Sciences. Diccionario Ingles-Espanol de termines de ganaderia, C91
International agencies directory, A65
International Association of Agricultural Librarians and Documentalists. Quarterly bulletin, G1
International catalogue of scientific literature ... 1901-1914, C8; F, Meteorology, D78; M, Botany, B20
International code of botanical nomenclature, International Congress of Botany. 8th, Stockholm, 1954, B75
International Congress of Botany. 8th, Stockholm, 1954. International code of botanical nomenclature, B75
International Committee on Zoological Nomenclature. Bulletin of zoological nomenclature, p. 71
International Committee on Zoological Nomenclature. Opinions and declarations, p. 71
International Council for the Exploration of the Sea. Journal du conseil, C473
International Council for the Study of the Sea. Index to publications by the International Council for the Exploration of the Sea, 1899-1938, C479
International directory of pedigree stock breeders, C96

International Federation for Documentation. Publication no. 247, A4
International Federation of the Agricultural Press. Repertoire international de la press agricole, A112
International Game Fish Association. Year book, C522
International green book of cottonseed and other vegetable oil products, E112
International Hydrographic Bur. Hydrographic dictionary, C507. Special publication no. 32, C507
International index to periodicals, p. 166
International Institute of Agriculture. Annuaire international de législation agricole, F23. Aperçu des bibliographies courantes concernant l'agriculture et les sciences connexes, V. A. Schaeffer, and others, A5. Atlas international zootechnique, C159. L'aviculture dans le monde, C214. Bibliographie d'agriculture tropicale, A27. Bibliographie internationale d'économie rurale, F3. Bibliothèques agricoles dans le monde et bibliothèques spécialisees dans les sujets se rapportant à l'agriculture, A69. Catalogue systématique [of the library], A30. L'enseignement agricole dans le monde, F40. Catalogue des periodiques possédés par la bibliothèque, A113. Decennial index of publications (1930-1939), A28. The first world agricultural census (1930), F16. Les institutions de génie rural dans le monde, D68. Les institutions d'expérimentation agricole dans les pays chauds, A71. Les institutions d'expérimentation agricole dans les pays tempérés, A70. Les institutions de zootechnie

International Institute
of Agriculture (continued)
dans le monde, C97. International trade of wood, statistical figures for the years 1925 to 1939, B281. International yearbook of agricultural statistics, F17. International yearbook of forestry statistics, B282. Liste chronologique des publications, A29. Recueil de coefficients et d'équivalences, A128. Système de classification des sciences agricoles, A134
International Institute of Refrigeration. Bulletin, E18
International review of poultry science, p. 84
International Seed Testing Association. Proceedings, B90
International Society of Soil Mechanics and Foundation Engineering. Technical terms in English, French, German, Portuguese, and Spanish used in soil mechanics and foundation engineering, D46
International Society of Sugar Cane Technologists. Memoir no. 1, C330
International Union of Biological Sciences. Index des généticiens, B153. Index des zoologistes, C48
International yearbook of agricultural statistics, F17
International yearbook of forestry statistics, International Institute of Agriculture, B282
Internationale zeitschrift für vitaminforschung, E19
Iowa. State College of Agriculture and Mechanic Arts. Library. Basic books and periodicals in home economics, E4
Irrigation
abstract journal, D54
bibliography, D59
Irrigation engineering, I. E. Hook, p. 136

Irrigation principles and practices, O. W. Israelson, p. 136
Irving, A. J. The directory of American horticulture for 1954, B117
Irwin, R. British bird books, C24
Ishida, M. The dictionary of terms used in entomology (English-Japanese), C337
Israelson, O. W. Irrigation principles and practices, p. 136

Jacks, G. V. Multilingual vocabulary of soil science, D47
Jackson, B. D. Glossary of botanic terms with their derivation and accent, B43. Guide to the literature of botany, B10. Index kewensis, B18
Jacobs, K. List of serials currently received in the Library of the United States Department of Agriculture, U.S.D.A. Bibliog. B. 12, A117. Livestock financing in the United States, C73
Jacobs, M. B. The chemical analysis of foods and food products, E140. The chemistry and technology of food and food products, E141
Jacoby, R. Das imker ABC: nachschlagewerk für alle gebiete der bienenzucht, C413
Jaeger, E. C. A dictionary of Greek and Latin, combining forms used in zoological names, C35. A source-book of biological names and terms, C36
Jahresbericht für agrikultur-chemie, D10
Jahrbuch für wissenschaftliche und praktische tierzucht, p. 74
Jahresbericht physiologie und experimentelle pharmakologie, p. 73
Jahresbericht über das gebiet der pflanzenkrankeiten ... , B172
Jahresbericht über die fortschritte der tierchemie, p. 73

Jahresbericht über die fortschritte in der untersuchung der lebensmittel (Nahrungs- und genussmittel, sowie gebrauchsgegenstände), E37
Jahresbericht über die leistungen auf dem gebiete der veterinär-medizin, p. 91
Jahresbericht veterinär-medizin, p. 91
Jardine, N. K. The dictionary of entomology, C338
Jasny, N., and others. Agricultural geography of Europe and the Near East, A100
Jeannel, R. G. La genèse des faunes terrestres, C54
Jeffers, F. H. A selection of old poultry books, C167
Jenkins, J. T. Bibliography of whaling, C480. A history of the whale fisheries, C540. The sea fisheries, C541
Jersey cattle, C133, C140-C141
Johansson, I. Förteckning över pälsdjurslitteratur, C25. Husdjursraserna: hästar, nötkreatur, får och getter, svin, kaniner och pälsdjur, fjäderfä, C84
Johnson, A. Bibliografiske hjelpemidler på landbrukets fagområder, A6
Jones, M. P. 4-H club insect manual, C398
Jordan, D. S. The genera of fishes ... , C508
Josephy, E. and Radt, F. Elsevier's encyclopaedia of organic chemistry, D15
Journal of dairy science, E20
Journal of home economics, E21
Journal of the Institute of brewing, Institute of Brewing, London, E17
Judah, C. B. The North American fisheries and British policy to 1713, C542
Judkins, J. National associations of the United States, A123
Jull, M. A. The races of domestic fowl, C192
Junk, W. Entomologen-adressbuch, C353

Just's botanischer jahresbericht, B21

Kahle, I. and Horn W. Über entomologische sammlungen, ein beitrag zur geschichte der entomo-museologie, C397
Kahn, H. Records in the National Archives relating to the range cattle industry, 1865-1895, C113
Kansas Agricultural Experiment Station. Department of Poultry Husbandry. Contribution no. 147, C168
Kansas State College, Manhattan. Department of Poultry Husbandry. Contribution no. 197, C175
Kaplan, L. Research material in the social sciences, p. 160, 166
Karny, H. Der insektenkörper und seine terminologie, C339
Kauffmann, F. Dictionnaire technique des matériels agricoles et de travaux publics, D64
Keefe, R. V. Sources of information concerning the commercial fisheries, C466
Keller, A. de. Elenchus librorum de apium cultura, C403
Kent, F. L., Stratton, G. B., and Smith, W. A., eds. World list of scientific periodicals published in the years 1900-1950, A122
Kew. Royal Botanic Gardens. Library. Catalogue, B22
Kingan & Co., Inc. A dictionary of packinghouse terms, E83
Kirkaldy, G. W. A bibliography of sugar-cane entomology, C327
Kirsten, G. Vollständiges wörterbuch der bienenkunde und bienenzucht, C420
Kittaku, R. Moderne tierärztliche terminologie, C254
Klages, K. H. W. Ecological crop geography, A103
Klee, R. Bibliotheca veterinaria, C233

Knight, H. L. Agricultural research institutions and library centers in foreign countries, A79
Knight, R. L. Dictionary of genetics, including terms used in cytology, animal breeding, and evolution, B149
Knöchel, E. Fachwörterbuch für die zuckerindustrie, E85
Köstler, J. Bibliographia forestalis, B202. Forstliche rundschau, B201
Kolloid zeitschrift, D23
Korschefsky, R. Verzeichnis der sammelindices von zeitschriften mit mehr oder weniger entomologischen inhalt, C391
Kracíková, L. Rusko-český slovník z oboru masnéhe prumyslu, E84
Kraemer, J. H. Wood conservation bibliography, B208
Kramer, O. Kellerwirtschaftliches lexikon, E86
Krancher, O. Kleines lexikon der bienenzucht und bienenkunde, unter teilweiser berücksichtigung von geschichte und pflanzenkunde für bienenzüchter, C421
Krasnosselskaja, T. C., Ussowsky, B. N., and Geminova, N. F. English-Russian agricultural dictionary, A60
Kreiger, L. C. C. Catalogue of the mycological library of Howard A. Kelley, B173
Krok, T. O. B. N. Bibliotheca botanica suecana ab antiquissimis temporibus ad finem anni mcmxviii, B23
Krüger, G. Der anatomische wortschatz unter mitberücksichtigung der histologie und der embryologie für die studierenden der veterinärmedizin, C255

Laignel-Lavastine, M. Historie générale de la médecine, de la pharmacie, de l'art dentaire et de l'art vétérinaire, C280
Le lait, E22
Land utilization bibliography, D35, G8
Landbouwdocumentatie, Netherlands. Ministerie van Landbouw, Visserij en Woedselvoorziening, A16
Landis, B. Y. Guide to the literature of rural life, F31
Landscape gardening see Horticulture
Landwirtschaftliche jahrbücher, p. 13
Landwirtschaftliches zentralblatt, A15; Abt. III, C65; Abt. 4, C220
Lanjouw, J., and others. International code of botanical nomenclature, B75
Lanjouw, J. and Stafleu, F. A. Index herbariorum, B50
Lauche, R. Internationales Handbuch der Bibliographien des Landbaus, A6a
Lawrence, G. H. M. Literature of taxonomic botany, B3
Lawrence, J. M. American livestock biographies, C114
Laws, Agricultural see Agriculture. Laws and legislation
Leavitt, C. T. Attempts to improve cattle breeds in the United States, 1790-1860 ... , C115
Leclainche, E. Histoire de la médecine vétérinaire, C281
Lee, G. A. and Penney, F. F. comps. Record group 114. Records of Soil Conservation Service; preliminary check list of records of Soil Conservation Service 1928-1942, D40
Legislation, Agricultural see Agriculture. Laws and legislation
Lemée, A. M. V. Dictionnaire descriptif et synonymique des genres de plantes phanérogames, B59
Leonard, L. L. International regulation of fisheries, C558
Leonard, M. D. Annotated bibliography of Puerto Rican entomology, C320

Leonard, W. H. and Martin, J. H. Principles of field crop production, B125
Lepidopterology dictionary, C336
Levi, W. M. The pigeon, C196
Lewis, E. A. Farm relief and Agricultural adjustment acts, G30. Federal farm loan act with amendments, and Farm mortgage and Farm credit acts, G31. Laws relating to agriculture, F25
Libraries see Agricultural library directory
Lichtenfelt, H. F. W. Literatur zur fischkunde, C474
Lincoln, W. American cookery books, 1742-1860, E38
Lindau, G. and Sydow, P. Thesaurus litteraturae mycologicae et lichenologicae ratione habita praecipue omnium quae adhuc scripta sunt de mycologia applicata quem congesserunt, B174
Lindley Library, The; catalogue of books, pamphlets, manuscripts, and drawings, Royal Horticultural Society, London. Library, B97
Lindstrom, D. E. American farmers' and rural organizations, A124
Lisbon. Escola Superior de Medicina Veterinaria. Indice bibliográfico dos escritos produzidos pelos autores veterinários portugueses, C234
Lissot, G. Dictionnaire d'aviculture, C188
List, R. J. Smithsonian meteorological tables, D84
List of publications of the Agriculture Department, 1862-1902, U. S. Superintendent of Documents, A44
List of serials currently received in the Library of the United States Department of Agriculture, November 1, 1949, U. S. Department of Agriculture. Library, A117

List of the existing entomological societies in the United States and Canada, C354
Litschauer, R. von. Vocabularium polyglottum vitae silvarum, B236
Little, E. L., Jr. Check list of native and naturalized trees of the United States (including Alaska), B261
Littré, É. Dictionnaire de médicine, de chirurgie, de pharmacie, de l'art vétérinaire et des sciences qui s'y rapportent, C256
Livestock
  bibliography, C83
  breeds, C84-C88, C101, C104-C105, C114, C116
  dictionaries, C90-C94
  diseases
    abstract journals, C218-C220
    bibliographies, C221-C242
    encyclopedias, C243-C248
    See also Veterinary medicine
  encyclopedias, C84-C88
  feeding, C102-C103
    abstract journals, E13
    bibliography, D27
  financing, C73
  geography, C111, C157-C163
  handbooks, manuals, textbooks, etc., C99-C105
  history, C106-C156
  marketing bibliography, C70
  physiology, C164
  statistics see Agricultural statistics
Livestock industry see Animal industry
Lloyd Library, Cincinnati, Ohio. Bibliographical contributions from the Lloyd Library, B24
Lockwood's directory of the paper and allied trades, B248
London. Royal College of Veterinary Medicine. The register of veterinary surgeons and the supplementary veterinary register, C268
London. Royal College of Veterinary Surgeons. A register of members of the Royal College of Veterinary Surgeons, from

London. Royal College of
Veterinary Surgeons (continued)
January 1794 to December
1851, inclusive, C267.
—— Library. Catalogue of the
historical collection: books
published before 1850, C227
London bibliograpphy of the social
sciences, p. 166
Lotsya, a biological miscellany,
v. 2, B149
Low, D. The breeds of domestic
animals of the British
Islands, C116
Low temperature research bibliography, B89, B134
Lowenstein, E., rev. American
cookery books, 1742-1860,
E38
Lübbert, H. Die grossen seefischereien der erde, ihre
faggründe und ihre fangprodukte, C527
Lumber trade
directories, B249-B250, B255,
B257
handbooks, manuals, texts,
B273
statistics, B278-B282
Lumberman's handbook and directory of the Southern forest
industries, 1952, B249
Lumberman's handbook and directory of the Western forest
industries, 1952, B250
Lyons, École National Vétérinarie.
Bibliothèque. Catalogue de
la bibliothèque, C228

Maatsch, R. Parey's illustriertes
gartenbaulexikon, G14
MacDonald, J. and Sinclair, J.
History of Aberdeen-Angus
cattle, C129. History of
Hereford cattle, C135
MacDonald, R. M. E. An analytical
subject bibliography of the
publications of the Bureau of
Fisheries, 1871-1920, C484
Mach, F., ed. Jahresbericht für
agrikultur-chemie, D10
Maciolek, J. A. Artificial fertilization of lakes and ponds; a
review of the literature, C481
McDonald, D. Agricultural writers
from Sir Walter of Henly to
Arthur Young, 1200-1800, A31
McFarland, R. History of the New
England fisheries, C543
McIlvain, Z. E. Publications on
planning for soil, water, and
wildlife conservation, flood
control, and land utilization,
D35
McIndoo, N. E. Plants of possible
insecticidal value, C438
McLester, J. S. Nutrition and diet
in health and disease, E142
Malin, D. F. The evolution of
breeds, C117
Mann, M. P. and Storer, T. I. Bibliography of rodent control,
C440
Manual of plant diseases, F. D.
Heald, p. 49, 50
Manual of sugar companies, E113
Manuale lactis, E39
Marketing bibliography, F13
Marr, E. B., Crane, E. J., and
Patterson, A. M. Guide to
the literature of chemistry,
D3
Marshall, M. L. and Doe, J. Handbook of medical library practice, C216
Marshall, W. T. and Woods, R. S.
Glossary of succulent plant
terms, B44
Martin, J. H. and Leonard, W. H.
Principles of field crop production, B125
Martiny, B. Wörterbuch der milchwirtschaft aller länder, E87
Marvel, T., ed. Frank Schoonmaker's
dictionary of wines, E91
Marzell, H. Wörterbuch der
Deutschen pflanzennamen, B45
Masera, E. La malattie infettive
degli insetti e loro indice
bibliografico, C320
Mason, I. L. World dictionary of
breeds, types, and varieties
of livestock, C85
Massachusetts Horticultural Society.
Library. Catalogue, B91

Matsuura, H. A bibliographic monograph on plant genetics (genic analysis) 1900-1929, B146
Matthes, E. Welche zeitschriften sind für ein zoologisches institut am wichtigsten, C59
Mattice, M. R. Bridges' food and beverage analyses, E143
Mauk, J. F. Industrial research laboratories of the United States, A72
Mayr, E. Methods and principles of systematic zoology, C45
Meat dictionary, E84
Meat industry
  directories, E97-E98, E127, E131
  encyclopedias, E88
  history, C109
  literature guide, E2
Meat inspection
  bibliography, C77
  directory, C98
Meat magazine. Meat packing cyclopedia, E114
Meat marketing bibliography, C70
Meat packing dictionary, E83
Meat packing industry
  bibliography, E31
  directories, E98, E114
Medical zoology bibliography, C237
Medical entomology bibliography, C219, C236
Medicine
  bibliographies, C9
  library handbooks, C216-C217
  See also Veterinary medicine
Meelig, M. L. Subject headings used in the Library of New York State College of Forestry, B284
Meisel, M. A bibliography of American natural history; the pioneer century, 1769-1865, B11
Melander, A. L. Source book of biological terms, C37
Melchior, H. and Werdermann, E., rev. Syllabus der pflanzen-familien, B14
Mellon, M. G. Chemical publications, their nature and use, D4
Menu terms, E75, E92
Merck & Co., Inc. Alpha tocopherol (vitamin E) annotated bibliography, E40. Annotated bibliography, penicillin in veterinary medicine, C241. Biotin; annotated bibliography, E41. Nicotinic acid; annotated bibliography, E42. Pantothenic acid; annotated bibliography, E43. Riboflavin; annotated bibliography, E44. Streptomycin and dihydrostreptomycin in veterinary medicine; annotated bibliography, C242. Vitamin B_1 (thiamine hydrochloride), annotated bibliography, E45. Vitamin B_6 (pyridoxin.:), annotated bibliography, E46. Vitamin B_{12}; selected annotated bibliography, 1954, E47. Vitamin K; annotated bibliography, E48
Merck index of chemicals and drugs, C455
Merillat, L. A. and Campbell, D. M. Veterinary military history of the United States, C282
Merrill, E. D. and Walker, E. H. A bibliography of Eastern Asiatic botany, B25
Metcalf, C. L. and Flint, W. P. Destructive and useful insects, C370
Meteorological abstracts and bibliography, American Meteorological Society, D76
Meteorology
  abstract journals, D76
  atlases, A104-A105
  bibliographies and indexes, D77-D80, G8, G22
  dictionaries, D81-D82
  handbooks, manuals, textbooks, D83-D85
Meyer, A. B. and Forbes, R. D., eds. Forestry handbook, B270
Meyer, H. Buch der holznamen, B237
Michaelis, A. and Rieger, R. Genetisches und cytogenetisches

Michaelis, A. and Rieger, R. (cont'd.)
 wörterbuch, B150
Milan. R. Scuola Superiore di
 Medicina Veterinaria. Biblioteca. Catalogo, C229
Milchwirtschaftliche forschungen, E23
Milk and milk products
 abstract journals, E20, E22-E23
 bibliographies and indexes, E39, E54, E61
 dictionary, E87
Millard's farm equipment directory, D69
Miller, C. G., Memorial Apiculture Library catalog, C411
Miller, T. L. History of Hereford cattle ... , C136
Miller, W. C. and West, G. P. Black's veterinary dictionary, C257
Millet, B136
Milling see Flour and feed trade
Milum, V. G. Information on beekeeping journals, C433
Mineral deficiencies in farm animals bibliography, C75
Mites and ticks, C329
Modern brewery age blue book issue, 1954, E115
Moll, L. La connaissance générale du boeuf, C118
Molon, G. Bibliografia orticola, B92
Monthly catalog of United States government publications, U. S. Superintendent of Documents, p. 103
Monthly checklist of state publications, U. S. Library of Congress, p. 103
Moore, H. A union checklist of forestry serials, B277
Moore, S. A. and H. S. History and law of fisheries, C544
Mormon cricket control directory, C360
Morrison, F. B. Feeds and feeding, C102
Morse, E. W. The ancestry of domesticated cattle, C119
Mortimer, C. F. Fertilizers in fishponds, C482
Mosquito news, p. 99, 100
Motor fuels bibliography, D63
Mrak, E. M. and Stewart, G. F. Advances in food research, E1
Mrak, E. M. and Shank, R. E. A bookshelf on foods and nutrition, G24
Mrugowski, H. Fachwörterbuch für die holzwirtschaft, B238
Müller, C. H. Lexicon entomologicum, oder entomologisches wörterbuch, C340
Müller, J. Terminologia entomologica, C341
Müller, K. Weinbau-lexikon, E87
Muenscher, W. C. L. Poisonous plants of the United States, B76. Weeds, B77
Mulder Bosgoed, D. Bibliotheca ichthyologica et piscatoria, C476
Munns, E. N. A selected bibliography of North American forestry, B209
Musgrave, A. Bibliography of Australian entomology, 1775-1930, C321
Musk-ox bibliography, C21
Mycologia ... , B175
Mycology see Fungi

National Canners' Association. Canners' directory, 1954, E116
National Cooperative Project. Conservation of nutritive value of foods. Beef, an annotated bibliography, E49
National Council on Hotel and Restaurant Education. Food preparation and related subjects, E50
National farm and tractor implement blue book, D71
National fruit and vegetable directory, 1951-1952, B118
National Geographic Society. Book of fishes, C528
National Hay Association. Year book, 1952-1953, B119

National Institute of Agricultural Engineering (Gt. Brit.). Agricultural and horticultural engineering abstracts, D55
National Lumber Manufacturers' Association. Lumber industry facts, B268
National Phosphorus Research Work Group. Summary of phosphorus research in the United States relating to soils and fertilizers, D32
National Poultry Improvement Plan directory, C200
National Research Council. Bulletin 111, E114. Bulletin 115, A125. Handbook of scientific and technical societies and institutions of the United States and Canada, A125. Industrial research laboratories of the United States, A72. N.R.C. 1059, C74. Publication 241, C439; 379, A72
―――― Chemical-Biological Coordination Center. Summary tables of biological tests, C456
―――― Chemistry Subcommittee. Annotated bibliography of analytical methods for pesticides, C439
―――― Committee on Animal Nutrition. Recommended nutrient requirements for domestic animals, C103, C215
―――― Committee on Food Habits. Manual for the study of food habits, E144
―――― Committee on Survey of Food and Nutrition Research. Survey of food and nutrition research in the United States, E145
―――― Food and Nutrition Board. Tables of food composition giving proximate mineral and vitamin components of foods (on a 100 gram basis) based on N.R.C. November 1, 1943 Army tables, E156

―――― Highway Research Board. The use of agricultural soil maps and the status of agricultural soil mapping in the United States, D51
―――― Insect Control Committee. Report no. 182, C440
National Research Council, Canada. Bibliography on nutrition in sheep, C74. Bibliography on the influence of mineral deficiencies on growth and reproduction of farm animals, C75. Bibliography on wool, C76
National Retail Farm Equipment Association. Official tractor and farm equipment guide, July-December 1953, D70
National Turkey Improvement Plan directory, C201
National Wildlife Federation. Directory of organizations and officials concerned with the protection of wildlife and other natural resources, C49
National wildlife refuge bibliography, P. A. Dumont, C20
Die natürlichen pflanzenfamilien nebst ihren gattungen und wichtigeren arten insbesondere den nutzpflanzen, A. Engler and K. Prantl, B15
Naturalist's directory, C355
Neave, S. A. Nomenclator zoologicus, C38
Nederlandsche Entomologische Vereeniging. Catalogus der bibliotheek, C306
Nematodes, plant resistance to, bibliography, C246
Nestler, R. B. Bibliography on poultry husbandry, C179. Bibliography on poultry industry, C180
Netherlands. Departement van Landbouw, Visscherij en Woedselvoorziening. Afdeeling Documentatie. Mededelingen, G5. Landbouwdocumentatie, A16
Neumann, L. G. Biographies vétérinaires, C269

New York Agricultural Experiment Station, Ithaca. Extension publications on poultry science issued by the agricultural experiment stations of the United States ending December, 1942, C184

New York Botanical Garden. North American flora, B60. Taxonomic index, B26

New York Public Library. An annotated list of bibliographies of cookery books, E51. List of works in the New York Public Library relating to fishing and fish culture, C491. Regional American cookery, 1884-1934, E52. The Whitney cookery collection, E53

Newbigin, M. I. Plant and animal geography, C55 and C160

Newton, H. K. and Ahern, G. P. A bibliography on woods of the world, exclusive of the temperate region of North America and with emphasis on tropical woods, B203

Niemann, G. Zoologisches wörterbuch, C39

Niklas, H. Literatursammlung aus dem gesamtgebiet der agrikulturchemie, D13 and D33

Nissen, C. Schone fischbücher, C490

Nodier, C. Bibliographie entomologique, C300

Nomenclator zoologicus, S. A. Neave, C38

Nordisk jordbrugsforskning. Organ for nordiske jordbrugsforskeres forening, A17

Norman, A. G. Advances in agronomy, B82

Norman, J. R. History of fishes, C545

North American flora, New York Botanical Garden, B60

Northwest livestock directory, p. 77

Northwestern miller, Almanack number, 1953, E117. A list of flour mills in the United States and Canada, E118

Notes on research in home economics education, U. S. Office of Education, E66

Nouveau Larousse agricole, A49

Nowakowska, J. and Wiebe, R. The technical literature of agricultural motor fuels including physical and chemical properties, engine performance, economics, patents, and books, D63

Nurseries see Horticulture. Directories

Nut varieties, B112

Nutrition
  abstract journals, E13, E27
  dictionaries, E82
  handbooks, manuals, textbooks, E137, E142, E154
  research, E145
  See also Food

Nutrition abstracts and reviews, Commonwealth Bureau of Animal Nutrition, E13

Nutrition reviews, E5

Oats, B141

Oceanography
  bibliography, C471
  dictionary, C502

Office International du Vin. Annuaire, 1951, E146. Bulletin, E24

Official methods of analysis, Association of Official Agricultural Chemists, D18

Officials of the U. S. Department of Agriculture, U. S. Department of Agriculture, A77

Oklahoma livestock breeders directory, p. 77

Olmo, H. P. and Brooks, R. M. Register of new fruit and nut varieties, 1920-1950, B112

Opinions and declarations, International Committee on Zoological Nomenclature, p. 71

Organic chemistry see Agricultural chemistry

Ornithology see Birds

Osborn, H. A brief history of entomology, C380. Fragments of entomological history, C381

Ostertag, R. Bibliographie der fleischbeschau, C77

Ostvold, H. Bibliographic aids for literature research, G2

Otero, J. I. and Cook, M. T. A bibliography of mycology and phytopathology of Central and South America, Mexico, and the West Indies, B176. Partial bibliography of virus diseases of plants, B177

Ouchi, Y. Bibliographical introduction to the study of Chinese insects, C322

Oudemans, A. C. Kritisch historisch overzicht der acaralogie, C329

Oudemans, C. A. J. A. Enumeratio systematica fungorum, B194

Overbeck, J. A. Glossarium melitturgigum, C422

Oxford University. Imperial Forestry Institute. Current monthly record of forestry literature, B210

Painter, R. H. Insect resistance in crop plants, C371

Pan American Sanitary Bureau. List of medical and public health journals of Latin America, C288

Pan American Union. Division of Agricultural Cooperation. Tentative directory of agricultural periodicals, societies, experiment stations, and schools in Latin America, A114

Paper
  abstract journals, B199
  bibliography and indexes, B215, B225
  dictionaries, B226

Paper and pulp mill catalogue, 1949-1950, B251

Paper trade directories, B248, B251-B254

Paper yearbook, 1954, B252

Parasites see Helminthology

Pardo Garcia, L. Diccionario de ictiologia, piscicultura y pesca fluvial, incluyendo la legislación y la organización e historia del Servicio Piscícola, C509

Parey's illustriertes gartenbau-lexikon, G14

Parey's landwirtschafts-lexikon, A50

Paris. Institution National du Bois. Centre de Documentation. Buletin bibliographique, B211

Paris. Station Centrale d'Hydrobiologie Appliquee. Analyses bibliographiques, 1940-1950, C475

Parry, J. W. The spice handbook, E147

Pasture crops see Agronomy

Pastures bibliography, B94-B95

Patten, B. M. Early embryology of the chick, C208

Patterson, A. M. and Crane, E. J. Guide to the literature of chemistry, D3

Payne, L. F. List of poultry books, C168. Old poultry books, C169. Poultry theses, C175

Payne, L. F. and Avery, T. B. International poultry guide for flock selection, C193

Pellett, F. C. History of American beekeeping, C429

Penicillin, veterinary uses, bibliography, C241

Penna, C. V. Informe final, Reunión Técnica de Bibliotecarios Agrícolas de América Latina, A7

Penney, F. F. and Lee, G. A. comps. Record group 114. Records of Soil Conservation Service; preliminary check list of records of Soil Conservation Service 1928-1942, D40

Pennington, C. E. and Feldkamp, C. L. List of the publications on soils issued by the state agricultural experiment stations of the United States through 1926, D29

Pennsylvania Horticultural Society.
　　Library. Catalog, B93
Percheron, A. R. Bibliographie
　　entomologique, C302
Percheron horse, C154
Periodical bibliographies, A7,
　　A108-A122
Perkins, W. F. British and Irish
　　writers on agriculture, A32
Pest control
　　abstract journals, C434-C435
　　bibliographies, C436-C440
　　directories, C443-C445, C446,
　　　C457
　　handbooks, manuals, textbooks,
　　　C446-C460
　　laws and regulations, C461-
　　　C463
　　terminology, C441
　　See also Insecticides, Rodent
　　　control, Pesticides
Pest control abstracts, C435
Pesticide handbook, C457
Pesticides
　　bibliography, C439
　　handbooks, manuals, textbooks,
　　　C446-C460
　　laws and regulations, C446,
　　　C461-C463
Petch, C. E. Index to entomological publications of the Department of Agriculture,
　　1884-1936, C314
Peters, J. L. Check-list of birds
　　of the world, C197
Peterson, A. A manual of entomological techniques, C372
Peterson, C. E. Fishery statistics,
　　C554. Preliminary review
　　of the fisheries of the United
　　States, 1952, C529
Peterson, M. S. A guide to the
　　literature on food research
　　and technology, G23
Petrak, F. List of new species
　　and varieties of fungi, 1936-
　　1939, B191. Verzeichnis der
　　neuen arten, varietäten,
　　formen, namen und
　　wichtigsten synonyme der
　　pilze, 1932-1935, B191
Pfeffer, A. Bibliographie ento-
mologique de la Bohême, de la
　　Moravie, et de la Silésie pour
　　les années 1800-1940, C323
Phanerogams, B56-B59, B66
Pharmaceutical abstracts, p. 91
Pheasants, C194
Phillips, J. C. American game
　　mammals and birds, C26. A
　　natural history of the ducks,
　　C198
Phillips' paper trade directory of
　　the world, 1954, B253
Phin, J. A dictionary of practical
　　apiculture, C423
Phosphorus in soils bibliography,
　　D32
Peiters, A. J. A digest of pasture
　　research literature, B94.
　　A digest of some world pasture research literature, B95
Pigeons, C196
Pigs see Swine
Pinkett, H. T. Preliminary inventory of the records of the
　　Forest Service, B224
Pirán, A. A. Bibliografía entomologica Argentina, C324-
　　C325
Plant-breeding
　　abstract journals, B129-B132
　　bibliography and indexes, B130,
　　　B133-B148
　　dictionaries, B149-B150
　　directories, B151-B153
　　handbooks, manuals, textbooks,
　　　B154-B156
Plant breeding abstracts, Commonwealth Bureau of Plant Breeding and Genetics, B129
Plant disease reporter. U. S. Agricultural Research Service.
　　Horticultural Crops Research
　　Branch, B157
Plant diseases see Plant pathology
Plant genetics see Plant-breeding
Plant names, B35-B38, B41-B42,
　　B44-B46, B48, B102-B106,
　　B109
Plant nutrition
　　abstract journals, D25
　　bibliographies and indexes, D27,
　　　D33

Plant pathology
  abstract journals, B158-B164
  bibliography and indexes, B137-B138, B158-B180
  handbooks, manuals, textbooks, B180, B182-B188
  maps, B181
Plant pathology, Sir E. J. Butler, p. 50
Plant physiology see Botany. physiology
Plant products, see Botany, economic
Plant science literature; selected references comp. by the library staff, U. S. Bureau of Plant Industry. Library, B100
Plants, cultivated, B103
Plants, edible, B61-B66, B102
Plants, effect of temperature on, B89, B134
Plants see also Botany and its subdivisions
Plumb, C. S. A partial index to animal husbandry literature, C78. Registry books on farm animals, C79. Types and breeds of farm animals, C104
Poisonous plants of the United States, B76
Poland China swine, C151
Poljoprivredna bibliografija FNRJ, G6
Postell, W. D. An introduction to medical bibliography, C217
Post's paper mill directory, B254
Potassium in soils bibliography, D34
Poultry
  bibliography, C83, C166-C185
  breeds, C190-C198
  clubs, societies: directories, C202
  dictionaries, C188-C189
  diseases:bibliography, C238 see also Veterinary medicine and its subdivisions
  encyclopedias, C186-C187
  feeding, C215
  handbooks, manuals, textbooks, C206-C211

Poultry husbandry
  abstract journals, p. 155-156
  bibliographies, C166-C185
  world report, C214
Poultry industry
  bibliographies, C171-C172, C174, C180-C181
  directories, C199-C205, E133
  statistics, C212-C213
Poultry science, p. 85
Powell, F. W. The Bureau of Animal Industry, its history, activities, and organization, C120
Prantl, K. and Engler, A. Die natürlichen pflanzenfamilien nebst ihren gattungen und wichtigeren arten insbesondere den nutzpflanzen, B15
Prentice, E. P. The history of Channel Island cattle, Guernseys and Jerseys, C133
Prentice, E. P., and others. American dairy cattle, C121
Prescott, M. S. Holstein-Friesian history, C139
Pretoria. Veterinary Research Institute. Draft of an international list of animal diseases, C259
Pritzel, G. A. Thesaurus literaturae botanicae omnium gentium, inde a rerum botanicarum initiis ad nostra usque tempora, quindecim millia operum recensens, B12
Produce Reporter Company. Semiannual blue book, E119
Prose and poetry of the live stock industry of the United States, C165
Public Affairs Information Service. Bulletin, p. 77, 166
Puerto Rico. Departamento de Agricultura y Comercio. Glosario de biologia animal (español-inglés), C92. Glossary of animal biology (English-Spanish), C93. Monografia 5, B235
Puhlemann, K. Imkers wortschatz, C424
Pulp see Wood-pulp
Punnett, R. C. Notes on old poultry books, C170

Purves, C. M., and others. Agricultural geography of Europe and the Near East, A100

Quarterly cumulative index medicus, C9
Quick frozen foods (periodical). 1954 directory of frozen food processors of fruits, vegetables, seafoods, meats, poultry, concentrates, and prepared foods, E120

Raber, O. German-English dictionary for foresters, B239
Radcliffe, W. Fishing from the earliest times, C546
Radt, F. and Josephy, E. Elsevier's encyclopaedia of organic chemistry, D15
Range management
   bibliography, B87, C80
   glossary, B40
Range plant handbook, U. S. Forest Service, B78
Ransome, H. M. The sacred bee in ancient times and folklore, C430
Rasmussen, W. D. and Edwards, E. E. A bibliography on the agriculture of the American Indians, A85
Read, P. R. and Zimmerman, F. L. Numerical list of current publications of the United States Department of Agriculture, A46
Readers' guide to periodical literature, p. 3
Recipe books see Cook-books
Recommended nutrient allowances for domestic animals, National Research Council. Committee on Animal Nutrition, C103, C215
Record, S. J. and Hess, R. W. Timbers of the New World, B269

Red book, encyclopaedic directory of the wine and liquor industries, 1953, E121
Red tractor book, D72
Reed, H. S. A short history of the plant sciences, B71
Reese, A. M. Outlines of economic zoology, C62
Refrigerated storage warehouses directory, E129
Refrigeration abstract journals, E9, E18
Refrigeration abstracts, American Society of Refrigerating Engineers, E9
Registry books on farm animals, C. S. Plumb, C79
Rehder, A. Bibliography of cultivated trees and shrubs hardy in the cooler temperate regions of the Northern Hemisphere, B27. The Bradley bibliography, B28. Manual of cultivated trees and shrubs hardy in North America, exclusive of the subtropical and warmer temperate regions, B262
Reindeer bibliography, C21
Reitemeier, R. F. Bibliography of literature on the release and fixation of native and fixed nonexchangeable potassium and ammonium of soils and silicate minerals, D34
Renner, F. G. A selected bibliography on management of western ranges, livestock, and wildlife, C80
Repellents see Insecticides
Repertorium entomologicum, C296
Research material in the social sciences, L. Kaplan, p. 160, 166
Resumptio genetica, B130
Reunión Técnica de Bibliotecarios Agrícolas de América Latina, Turrialba, 1953. Informe, final, A7
Reuss, J. D. Repertorium commentationum a societatibus litterariis editarum, C15

Review of applied entomology.
Series A: Agricultural,
Commonwealth Institute of
Entomology, C293. Series
B: Medical and veterinary,
Commonwealth Institute of
Entomology, C219
Review of applied mycology, The.
Commonwealth Mycological
Institute, B160
Revue internationale des industries agricoles, A18
Rice, B142
Rice annual, 1954, B120
Rice trade directories, B120
Richardson, J. E. Conservation of nutritive value of foods. Beef, an annotated bibliography, E49
Richter, H., and others. German-English botanical terminology B34
Ridgeway, W. The origin and influence of the thoroughbred horse, C153
Rieger, R. and Michaelis, A. Genetisches und cytogenetisches wörterbuch, B150
Riley, C. V., bibliography, C312
Rinderpest bibliography, C239-C240
Robb, J. Notable angling literature, C492
Rockwell, J. E. Contents of index to bulletins of the Bureau of Plant Industry, nos. 1 to 100 inclusive, B96
Rodent control
  bibliography, C440
  laws and regulations, C462
Rogers, E. G. and McIlvain, Z. E. Publications on planning for soil, water and wildlife conservation, flood control, and land utilization, D35
Rojahn, C. A., ed. Jahresbericht über die fortschritte in der untersuchung der lebensmittel (Nahrungs- und genussmittel, sowie gebrauchsgegenstände), E37
Romanoff, A. L. Popular publications in the field of poultry industry, issued by the agricultural experiment stations of the United States ending June, 1941, C185
Romanoff, A. L. and A. J. The avian egg, C209
Root, A. I. and E. R. The ABC and XYZ of bee culture, C414
Rosa, H. Scientific and common names applied to tunas, mackerels, and spear fishes of the world ... , C510
Rose, A. and E. Distillation literature, index and abstracts, D11
Rothamsted Experimental Station. Library. Library catalogue of printed books and pamphlets on agriculture published between 1471 and 1840, A33
Rothschild, H. J. N. C. de. Bibliographia lactaria, B54
Rounsefell, G. A. Fishery science, C530
Royal Entomological Society of London. Catalogue of the library, C307. List of fellows, C356. List of serial publications in the Society's library, C392. Generic names of British insects, C345
Royal Horticultural Society, London. Library. The Lindley Library; catalogue of books, pamphlets, manuscripts, and drawings, B97
Royal Meteorological Society, London. Bibliography of meteorological literature, D79
Royal Society of London. Catalogue of scientific papers, 1800-1900, C16
R U S. A biographical register of rural leadership in the United States and Canada, 1918-1930, A91
Rural sociology
  abstract journals, F28-F29
  bibliographies and indexes, F30-F31
  directory, F32
  handbooks, manuals, textbooks, F33-F34
Rural sociology (periodical), F29
Rye, B143

Saccardo, P. A. Sylloge fungorum omnium hucusque cognitorum, B195

Sachs, J. von. History of botany, 1530-1860, B72

Salmon, D. E. Special report on the history and present condition of the sheep industry of the United States, C147. The United States Bureau of Animal Industry, at the close of the nineteenth century, 1884-1900, C122

Salmonsen, E. M. Bibliographical survey of vitamins, 1650-1930, E55

Sanders, A. H. The cattle of the world, C86. A history of Aberdeen-Angus cattle, C130. A history of the Percheron horse, C154. Red, white, and roan; stories relating to the origin of the Shorthorn breed of cattle in Great Britain; its early introduction into America, and its growth in popularity throughout the United States in recent years, C143. Shorthorn cattle, C144. The story of the Herefords, C137

Sanz Egaña, C. Enciclopedia de la carne, E88. Historia de la veterinaria española, C283

Sargent, C. S. Manual of the trees of North America (exclusive of Mexico), B263. The silva of North America, B264

Sauer, C. O. Agricultural origins and dispersals, C161

Say, T. A glossary to Say's Entomology, C342

Scattergood, L. W. Bibliographic sources for fishery students and biologists, C465. English translations of fishery literature, C496

Schaeffer, V. A., and others. Aperçu des bibliographies courantes concernant l'agriculture et les sciences connexes, International Institute of Agriculture, A5

Schenk, E. T. Procedure in taxonomy, C46

Schenkling, S. and Horn, W. Index litteraturae entomologicae. Ser. 1: Die welt-literatur über die gesamte entomologie bis inklusive 1863, C300

Schmaltz, R. Entwicklungsgeschichte des tierärztlichen berufes und standes in Deutschland, C284

Schmid, A. Rassenkunde des rindes, C87

Schmidt, L. B. Topical studies and references on the economic history of American agriculture, A92

Schmucker, T. The tree species of the northern temperate zone and their distribution, B265

Schneider, M. Agrarmeteorologische bibliographie, 1950, G22

Schoellhorn, F. Bibliographie des brauwesens, E56

Scholl, A. W. and Cartwright, D. R. Bibliography of synthetic plant-growth substances, G13

Schoonmaker, F. Frank Schoonmaker's dictionary of wines, E91

Schrader, G. W. Biographisch-literarisches lexicon der thierärzte aller zeiten und länder sowie der naturforscher, aertze, landwirthe, stallmeister, u.s.w., welche sich um die thierheilkunde verdient gemacht haben, C270

Schriftum der bodenkultur, Das. A19

Schroeder, C. W. M. Handbuch der entomologie, C373

Schulz, M. E. and Sydow, G. Manuale lactis, E39

Scottish Beekeepers' Association. Catalogue of the Moir Library, C405

Seed plants see Phanerogams

Seed trade directories. B116, B121, B124

Seed trade buyers' guide, 1954, B121

Seed bibliographies, B88, B90

Seiden, R. Livestock health encyclopedia, C244. Poultry handbook, C187

Sel'skokhoziaĭstvennaia literatura SSSR, A34

La semana vitivinicola. Directorio de la produccion, comercio y exportacion de vinos, alcoholes, vinagres, licores y de las industrias conexas, asi como de las principales cases extranjeras importadoras de vinos y espirituosos, E122

Senet, A. Histoire de la médecine vétérinaire, C285

Sewell, A. J. The dog's medical dictionary, C260

Seymour, A. B. Host index of the fungi of North America, B196

Shafer, J. A. and Britton, N. L. North American trees, being descriptions and illustrations of the trees growing independently of cultivation in North America, north of Mexico, and the West Indies, B258

Shank, R. E. and Mrak, E. M. A bookshelf on foods and nutrition, G24

Sheep
  bibliography, C71
  feeding, C74
  history, C146-C150

Shepard, H. H. The chemistry and action of insecticides, C458

Shepard, S. Bibliography of current horticultural periodicals, 1954, B128

Shepard, W. Spanish terms for forest rangers, B240

Sherman, J. D. Some entomological and other bibliographies, C291

Shibata, S. Saishin hikkei chikusan soten, C94

Shirai, T. History of Japanese veterinary medicine, C286

Shorthorn cattle, C142-C145

Shrubs, B79, B262
  bibliography, B27-B28

Silbermann, G. Énumération des entomologistes vivans, suivie de notes sur les collections entomologiques des principaux musées d'histoire naturelle d'Europe, sur les sociétés d'entomologie, sur les recueils périodiques consacrés a l'étude des insectes, et d'une table alphabétique des résidences des entomologistes, C357

Silva see Trees

Silva of North America, C. S. Sargent, B264

Silvae orbis no. 1, B276

Simon, A. L. Bibliotheca Bacchica, E57. Bibliotheca gastronomica, E58. Bibliotheca vinaria, E59. A concise encyclopaedia of gastronomy, E89. A dictionary of wine, E90

Sinclair, J. History of polled Aberdeen or Angus cattle, C131. History of shorthorn cattle, C145

Sinclair, J. and MacDonald, J. History of Aberdeen-Angus cattle, C129. History of Hereford cattle, C135

Skinner, H. The entomologists' directory, containing the names, addresses, special departments of study, etc., of those interested in the study of insect life in the United States and Canada, C358

Smallwood, N. W. Commercial horticultural organizations of the United States and Canada, B122. Horticultural organiations of the United States and Canada, B123

Smiley, E. M. and Artschwager, E. F. Dictionary of botanical equivalents, B33

Smith, A. D. B. Genetics of cattle, C81

Smith, C. L. Guide to the laws affecting insecticides, fungicides, and related products, C461

Smith, F. The early history of veterinary literature and its British development, C287

Smith H. The master dictionary of

Smith, H. (continued)
    food and cookery and menu translator, E92
Smith, J. B. Explanation of terms used in entomology, C343 and C332
Smith, L. M. and Bailey, S. F. Handbook of agricultural pest control, C447
Smith, R. C. Guide to the literature of the zoological sciences, C1
Smith, W. A., Kent, F. L., and Stratton, G. B., eds. World list of scientific periodicals published in the years 1900-1950, A122
Smithsonian Institution. Smithsonian meteorological tables, D84
Smithsonian Institution. U. S. National Museum. Bulletin 50, Part X, C195. Bulletin 67, C393
Smithsonian miscellaneous collection v. 114, D84
Smut fungi, The. G. W. Fischer, B193
Snedecor, G. W. Statistical methods applied to experiments in agriculture and botany, A129
Snell, W. H. Three thousand mycological terms, B197
Soap and chemical specialties. Blue book and catalog section, C445
Social insects, glossary, C332
Social science abstracts, p. 166
Société Centrale d'Apiculture et d'Insectologie. Catalogue de la bibliothèque ... , C406
Société d'Apiculture d'Alsace-Lorraine. Katalog der bücher und schriften der central-bibliothek ... , C407
Société Mycologique de France. Bulletin trimestriel, B162
Société Romande d'Apiculture. Bibliothèque. Catalogue, C408
Societies see Agricultural society directories

Society of American Bacteriologists. Bergey's manual of determinative bacteriology, B185
Society of American Foresters. Forestry handbook, B270. Forestry terminology, a glossary of technical terms used in forestry, B241
Soil conservation
    abstract journals, D24
    bibliographies and indexes, D30, D35, D38, D41-D42
    dictionaries, D44
    directories, D49
Soil conservation literature, U. S. Soil Conservation Service. Library, D41
Soil erosion bibliographies and indexes, D30, D43
Soil science (periodical), p. 13, 131
Soil survey manual, U. S. Bureau of Plant Industry, Soils, and Agricultural Engineering, D52
Soils
    abstract journals, A11, D8, D22-D25
    bibliographies and indexes, A23-A24, D26-D43
    colloidal aspects of: bibliography, D23
    dictionaries, D44-D47
    handbooks, manuals, textbooks, D51-D53
    maps, D51
    physical properties: bibliography, D26
Soils and fertilizers, Commonwealth Bureau of Soil Science, D22
Soils and men, U. S. Department of Agriculture, D53
Sorauer, P. Handbuch der pflanzenkrankheiten, B186
Sorghum, B136
Sorokin, P. A., Zimmerman, C. C., and Galpin, C. J. A systematic source book in rural sociology, F33
Sotham, W. H. History of the Herefords in America, C136
Soule, B. A. Library guide for the chemist, D5

Sources of business information, E. T. Coman, p. 160
Sources of engineering information, B. Dalton, p. 136
South African fishing industry handbook and buyers' guide, 1954, C521
Soybean blue book, 1954, E123
Soybean industry directory, E123
Spain. Ministerio de Agricultura. Boletín bibliográfico agricola, A20
Spanish terms for forest rangers, W. Shepard, B240
Spermatophyta see Phanerogams
Sport fishery abstracts, U. S. Fish and Wildlife Service, G18
Spices
  dictionary, B46
  handbooks, manuals, textbooks, E147
Stafleu, F. A. and Lanjouw, J. Index herbariorum, B50
Standardized plant names, American Joint Committee on Horticultural Nomenclature, B102
Stansby, M. E. Thirty-five year index of the fishery technological service publications of interest to fishery technologists, G20
Stapledon, R. G., ed. Farming and mechanised agriculture, A80
Stark, L. M. The Whitney cookery collection, E53
State agriculture departments directory, A75
State extension apiculturists directory, C362
State extension entomologists directory, C362
State game department bulletins bibliography, C22
Statistics, F14-F21. See also subheading statistics under various subjects
Stechow, M. Register der weltliteratur über vitamine ... , E60
Steffek, E. F. Plant buyers' guide of seed and plant materials in the trade, B124

Steinmetz, E. F. Codex vegetabilis, B46. Vocabularium botanicum, B47
Stephany, C. D. and Von Loeseke, H. W. Selected references on yeast, G26
Sterne, F. and Weil, B. H. Literature search on the preservation of food by freezing, E71
Stevens, F. C. Books of reference in zoology, chiefly bibliographical, C2
Stevenson, J. A. Foreign plant diseases ... a manual of economic plant diseases which are new to or not widely distributed in the United States, B187
Stewart, G. F. and Mrak, E. M. Advances in food research, E1
Stockholm. Veterinärhögskolan. Katalog, C230
Stockyard directory, C98
Storer, J. The wild white cattle of Great Britain, C123
Storer, T. I. and Mann, M. P. Bibliography of rodent control, C440
Stowage red book, A130
Stratton, G. B., Smith, W. A., and Kent, F. L., eds. World list of scientific periodicals published in the years 1900-1950, A122
Streptomycin, veterinary uses, bibliography, C242
Strickland, H. E., ed. Bibliographia zoologiae et geologiae, C11
Strobel, D. R., Bryan, W. G., and Babcock, C. J. Flavors of milk, a review of literature, E61
Strömberg, R. Studien zur etymologie und bildung der griechischen fischnamen, C511
Strong, R. M. A bibliography of birds, C166
Stubbe, H. Genetisch-pflanzenzüchterische bibliographie, 1939-1946 (1947), B147
Stud books, C79
Stuntz, S. C. List of agricultural periodicals of the United States and Canada published

Stuntz, S. C. (continued)
    during the century July 1810
    to July 1910, A115
Sturkie, P. D. Avian physiology,
    C210
Sturtevant, E. L. Sturtevant's
    notes on edible plants, B65
Succulent plant glossary, B44
Sugar-cane entomology, C327,
    C330
Sugar industry
    dictionaries, E85
    directories, E113, E124
Sugar reference book, 1953, E124
Summary tables of biological
    tests, National Research
    Council. Chemical-
    Biological Coordination
    Center, C456
Supreme Commander for the
    Allied Powers. Natural Re-
    sources Section. Report no.
    63, C504
Suttie, D. F. Dictionary of poultry,
    C189
Swain, R. B. Insect guide, C374
Swedish-English vocabulary for
    foresters, J. L. Deen, A. B.
    Benson, and J. J. Dannfelt,
    B234
Swine
    bibliography, C67
    history, C151-C152
Sydow, G. and Schulz, M. E.
    Manuale lactis, E39
Sydow, H. Annales mycologici
    editi in notitiam scientiae
    mycologicae universalis,
    B159
Sydow, P. and Lindau, G.
    Thesaurus litteraturae my-
    cologicae et lichenologicae
    ratione habita praecipue
    omnium quae adhuc scripta
    sunt de mycologia applicata
    quem congresserunt. B174
Systematic zoology see Zoology,
    Economic. Nomenclature

Taber, C. W. and Graves, L. G.
    A dictionary of food and
    nutrition, E82

Taxonomic index, New York
    Botanical Garden, B26
Taxonomy see Zoology, Economic.
    Nomenclature, and subhead
    Classification under specific
    subjects
Taxonomy of vascular plants, G. H.
    M. Lawrence, B3
Taylor, C. C., and others. Rural
    life in the United States, F34
Taylor, C. M. Food values in shares
    and weights, E157
Taylor, L. W. Fertility and hatch-
    ability of chicken and turkey
    eggs, C211
Taylor's encyclopedia of gardening,
    horticulture, and landscape
    design, B109
Tea trade directory, E126
Theiler, J. Katalog der bibliothek
    des Vereins deutsch-
    schweizerischer bienen-
    freunde, mit nachtrag 1941,
    C409
Thiessen, A. H. Weather glossary,
    D82
Thomas' register of American
    manufacturers, D73
Thomas' wholesale grocery and
    kindred trades register, E125
Thomazi, A. A. Histoire de la
    pêche des âges de la pierre
    a nos jours, C547
Thompson, D. W. Glossary of
    Greek fishes, C512
Thompson, E. K. and Gray, L. C.
    History of agriculture in the
    Southern United States to
    1860, A88
Thompson, G. F. Index to literature
    relating to animal industry in
    the publications of the U. S.
    Department of Agriculture,
    1837-1898, C82
Thompson, J. W. A history of live-
    stock raising in the United
    States, 1607-1860, C124
Thornthwaite, C. W. Atlas of
    climatic types in the United
    States, 1900-1939, A104
Thorpe, J. F. and Whiteley, M. A.
    Thorpe's dictionary of applied
    chemistry, D17

Timber see Wood
Timberman. Forest products industry directory of Western North America, 1950, B255
Tobacco, B145
Todd, E. F. H. Diccionario técnico-agrícola; ingles-español, G12
Torrey Botanical Club. Bulletin, B29. Index to American botanical literature, B29
Tower, W. S. A history of the American whale fishery, C548
Towne, C. W. and Wentworth, E. N. Pigs, C152. Shepherd's empire, C148
Tractor field book with power farm equipment specifications, D74
Tractors see Agricultural engineering
Trees, B79, B258-B266, B271 bibliography, B27-B28, B205, B221
Tressler, D. K. Marine products of commerce ... , C531
Trolinger, C. and Hawks, E. B. List of the publications on soils issued by the U. S. Department of Agriculture, 1844-1926, D31
Tropical agriculture bibliography, A27
Tropical beef cattle industry in the Western Hemisphere, C. Bien, C66
Tropical Plant Research Foundation. A bibliography on woods of the world ... with emphasis on tropical woods, B203
Trout fly dictionary, C523
True, A. C. A history of agricultural education in the United States, 1785-1925, F37. A history of agricultural experimentation and research in the United States, 1607-1925, including a history of the United States Department of Agriculture, A93. A history of agricultural extension work in the United States, 1785-1923, F38
Tucker, E. M. Catalogue, Harvard University. Arnold Arboretum. Library, B17
Turkeys see Poultry
Turkey world, p. 88
Turner, D. B. Professional opportunities in the wildlife field, C523
Turrialba, Suplemento bibliografico, A21
Two blades of grass, T. S. Harding, A89

Uebele, G. Handlexicon der tierärztlichen praxis, C245
Ukers' international tea and coffee buyers' guide, E126
Underwood, M. H. Bibliography of North American minor natural history serials in the University of Michigan libraries, C60
United Nations Educational, Scientific and Cultural Organization. Bibliography of interlingual scientific and technical dictionaries, A53. Directory of institutions engaged in arid zone research, A73. Directory of international scientific organizations, A126. Occasional papers in education no. 15, F35. Publication no. 863, A4. Teaching agriculture: a selected bibliography, F35
U. S. Agricultural Marketing Service. AMS-48, F5. Check list of reports ... , F4. Directory, names and addresses of personnel who may be contacted for grading and/or inspection service on poultry and egg products and domestic rabbits, C204. Directory of personnel in rural sociology, F32. Directory of state departments of agriculture, 1955, A75. Periodic reports, F5. Wholesale meat trade directory, E127

Index 221

U. S. Agricultural Research Service. Abstracts of recent published material on soil and water conservation, D24. A. P. H. 105, C201. ARS 53-1, C199. ARS 53-6, C200. Chemicals evaluated as insecticides and repellents at Orlando, Fla., C459. Directory of U.S.R.O.P. breeders qualifying for the U. S. Register of Merit Breeding Stage, 1952-1953, C199. Hatcheries and dealers participating in the National Poultry Improvement Plan, C200. Hatcheries, dealers, and independent flocks participating in the National Turkey Improvement Plan, 1954, C201. Plant disease reporter, B157. Publications [Production Economics Research Branch] issued from December [1953] through April [1954], F6. Published and processed reports of research in foods, human nutrition, and home economics at the land-grant institutions, E62. Regulations for the enforcement of the Federal insecticide, fungicide, and rodenticide act, C462. Service and regulatory announcements 166, C462. Working reference of livestock regulatory establishments, stations, and officials, C98

——— Armed Forces Medical Library. Current list of medical literature, C9, p. 90

——— Army Medical Library. Index-catalogue of the library of the Surgeon General's Office, United States Army ... authors and subjects, C9

——— Bureau of Agricultural Chemistry and Engineering. Current literature in agricultural engineering, D61. Poultry housing. C176

——— Bureau of Agricultural Economics. Agricultural economic and statistical publications, F7. Agricultural economics bibliographies, F10; 24, C171; 62, C73. Agricultural history series 5, C124. Available printed and processed publications issued by the Bureau of Agricultural Economics from July 1, 1940, to February 28, 1941, F8. A chronology of American agriculture, 1790-1940, A94. Economic library lists, F11; 4, C177. Egg auctions, C177. A graphic summary of farm animals and animal products (based largely on the census of 1940), C162. Printed publications issued by the Bureau of Agricultural Economics, F9

——— Bureau of Animal Industry. AHD 4, p. 77. AHD 6, p. 77. AHD 18, C202. Annual report, 27th, 1910, C119. Index-catalogue of medical and veterinary zoology, C237. Poultry specialty clubs and national associations, C202. Special report on diseases of cattle, C246. Special report on diseases of the horse, C247 history, C112, C120, C122

——— Bureau of Dairy Industry. BDI-Inf series 4, p. 77. BDI-Inf series 6, p. 77

——— Bureau of Chemistry and Soils. Index of publications ... 1862-1937, D36. Index of publications of the Bureau of chemistry and soils, originally the Bureau of chemistry and the Bureau of soils, D14

——— Bureau of Entomology. Bibliography of the more important contributions to American economic entomology, C312. Bulletin 24, n.s., C310. Bulletin 81, C309. List of entomological publications of personel of Cereal and Forage

222   Index

U. S. Bureau of Entomology (cont'd.)
Insect Investigations, U. S. Bur. of Entomology, 1904-1928, inclusive, C316 history, C402
—— Bureau of Entomology and Plant Quarantine. Directory ... 1935-1940, C359. E-297, C426. E-661, C438. Entomology current literature, C297. Information about bee culture, C415. List of dealers in beekeeping supplies, package bees, and queens, C426. List of entomological technique (ET-series) publications, C317. Some of the more important institutions conducting research in entomology; also the more significant entomological societies, C361. State leaders, grasshopper and Mormon cricket control, C360
—— Bureau of Fisheries. An analytical subject bibliography of the publications of the Bureau of Fisheries, 1871-1920, C484. Document 899, C484. Report of the Cmnr. ... 1919/20, C484. The United States Bureau of Fisheries, its establishment, functions, organization, resources, operations, and achievements, C550
—— Bureau of Foreign and Domestic Commerce. American lumber industry, B273. Cattle and dairy farming, C88
—— Bureau of Human Nutrition and Home Economics. Tables of food composition in terms of eleven nutrients, E158. Titles of completed theses in home economics and related fields in colleges and universities of the United States, E63
—— Bureau of Plant Industry. Agronomy current literature, B98. Botany current literature, B30. Bulletin 101, B96. Check list of publications issued by the Bureau ... 1901-1920, and by the divisions and offices which combined to form this Bureau, 1862-1901, B99. Check list of publications of the state agricultural experiment stations on the subject of plant pathology, 1876-1920, B178. A check list of the publications of the Department of Agriculture on the subject of plant pathology, 1837-1918, B179. Plant science literature, B100
—— Bureau of Plant Industry, Soils, and Agricultural Engineering. Bibliography of weed investigations, B31. Soil survey manual, D52. Special publication 1, B180
—— Bureau of Soils, bibliography, D14
—— Bureau of the Census. Fifteenth census of the United States: 1930. Agriculture. Chickens and chicken eggs and turkeys, ducks, and geese raised on farms. Chickens and poultry products, with selected items by size of flock, for the United States, states, and counties, 1930 and 1929, C212. Fisheries of the United States, C549. Sixteenth census of the United States: 1940. Agriculture. Special poultry report, C213. United States census of agriculture: 1950, F18
—— Congress. House. Committee on Agriculture. Research and related services in the United States Department of Agriculture, A95
—— Department of Agriculture. Agricultural history series 2, A96. Agricultural statistics, F19. Agriculture handbooks: 18, D52; 41, B261; 58, B80; 65, C398; 69, C459; 76, A76;

U. S. Department of
Agriculture (continued)
78, A78; 79, G29. Atlas of American agriculture, A105. Bibliographical bulletins, A35; 6, E64; 8, C437; 10, D63; 12, A117; 14, B167; 15, C70; 19, C66; 20, B205; 23, B52; 24, D57. Bibliography of animal industry literature, C82. Bibliography of marketing publications, F13. Bibliography of poultry publications, C178. Check list of the publications...on the subject of plant pathology, 1837-1918, B179. Climate and man, D85. Compilation of statutes relating to soil conservation, marketing quotas and allotments, crop insurance, sugar payments and quotas, price support, Commodity Credit Corporation, and related statutes as of January 1, 1955, G29. Department bulletin 1199, A43. Directory of organization and field activities of the Department of Agriculture: 1955, A76. Farmers' bulletin 619, p. 78; 767, p. 87; 1391, p. 87; 1443, p. 78; 1779, p. 78; 2065, p. 87; 2066, p. 87. Geography of the world's agriculture, A106. Index to publications, 1901-1940, A40. Insects, C363. Keeping livestock healthy, C248. List by titles of publications...from 1840 to June 1901, supplements to Dec. 1945, A41. List of available publications, A26. List of publications...1862-1902, with analytical index, A44. List of publications on soils, D31. Miscellaneous publications: 15, F38; 36, F37; 60, A26; 84, A86; 105, C157; 110, B40; 118, D77; 164, B148; 200, B74; 220, C359; 245, A116; 251, A93; 269, C158; 281, C80; 300, p. 87; 303, B79; 312, D30; 337, A108; 364, B209; 398, A115; 401, B53; 421, A104; 447, A85; 450, A46; 515, F27; 530, C152; 555, p. 87; 572, E158; 628, E155; 665, A100; 705, A102. Numerical list of current publications, A46. Officials of the U. S. Department of Agriculture, A77. Soils and men, D53. Some landmarks in the history of the Department of Agriculture, A96. Trees, B271. Weights, measures, and conversion factors, A131. Workers in subjects pertaining to agriculture in land-grant colleges and experiment stations, 1954-1955, A78. Yearbook of agriculture, A81; 1938, D53; 1941, D85; 1942, C248; 1949, B271; 1952, C363, C454

biography, A97
directories, A76-A77
entomological publications: bibliography, C315, C319
history, A89, A93, A96, A98-A99
laws and legislation, F26
—— —— Division of Entomology. Bulletin 19, C313. An enumeration of the published synopses, catalogues, and lists of North American insects, C313
—— —— Division of publications. Bulletins 5, C82; 6, F12
—— —— Library. Agricultural economics literature, F12. Agricultural library notes, A36. Bibliographical contributions, A37; 1, B179; 2, B179; 3, B99; 13, D37; 14, D31; 15, D29; 19, C71; 20, C315; 21, C403. Bibliography of agriculture, A24. Bibliography on dehydration of foods, 1938-1943, E64. Bulletin: 18, C173; 55, C308. Catalogue of the publications relating to entomology in the library of the U. S. Dept. of Agriculture, C308. Classification scheme of the Bibliography of Agriculture,

U. S. Department of Agriculture. Library (continued)
January 1950, A135. A classified list of soil publications of the United States and Canada, D37. Cotton literature, B101. Library lists, A38; 1, A39; 12, C174; 16, B123; 17, B122; 18, A118; 30, A135. List of periodicals currently received in the Library of the United States Department of Agriculture, June 1, 1936, A116; Nov. 1, 1949, A117. Partial list of United States farm papers received in the U. S. Department of Agriculture Library (state and county farm bureau and state college papers not included), A118. Scheme of classification for the United States Department of Agriculture Library, A136. Selected list of American agricultural books and periodicals, A39. Soil conservation bibliographies, D38. Soil conservation service library lists, D39.

────── Office of Information. Abridged list of federal laws applicable to agriculture, F24. Biographies of persons in charge of federal agricultural work, 1836 to date, A97. Condensed history of the U. S. Department of Agriculture, A98. Document 1, A99; 2, F24; 3, A97; 4, A98. Index to the publications of the United States Department of Agriculture, 1901-1925, A40. List by titles of publications of the United States Department of Agriculture from 1840 to June 1901, A41. Origin, structure, and functions of the U. S. Department of Agriculture, A99

────── Department of Commerce. Basic information sources on agricultural machinery and implements, D62. Business service bulletin 37, D62. Food processing, marketing, and equipment (Basic information sources), E65

────── Extension Service. List of State extension entomologists and extension apiculturists, C362

────── Fish and Wildlife Service. Available leaflets on fisheries, C485. Circular 36, G19. Commercial fisheries abstracts, C467. Doctoral dissertations on the management and ecology of fisheries, C483. Fishery leaflet 9, C485; 160, C553; 195, p. 123; 197, C554; 254, p. 123; 255, C449; 292, p. 123; 293, p. 123; 313, p. 123; 359, p. 123; 362, C466; 393, C529. Fishery motion pictures, C499. Fishery publication index, 1920-1954, G19. Fishery resources of the United States, C532. Fishery statistics of the United States ... , C555. Monthly list of printed and duplicated material, C486. Partial list of journals and newspapers concerning the fisheries, C553. Sources of information concerning the commercial fisheries, C466. Special scientific report: fisheries 31, C483; 35, C496; 113, C481; 127, C477; 176, C496. Sport fishery abstracts, G18. Thirty-five year index of the fishery technological service publications of interest to fishery technologists, G20. Wildlife abstracts, 1935-51, C5. Wildlife leaflet, 334, C20. Wildlife review, C4

────── Foreign Agricultural Service. Published information on agriculture in foreign countries, January 1937-December 1953, A42

────── Forest Service. Check list of the native and naturalized trees of the United States

U. S. Forest Service (continued)
including Alaska, B266.
Lumber manufacturing, wood
using, and allied associations,
B257. Organizational directory ... March 1953, B256.
Range plant handbook, B78
────── ────── Forest Products Laboratory. List of publications
on chemistry of wood and
derived products, B212. List
of publications on logging,
milling, and utilization of
timber products, B213. List
of publications on mechanical properties and structural
uses of wood and wood products, B214. List of publications on pulp and paper, B215.
List of publications on the
growth, structure, and identification of wood, B216. List
of publications on the seasoning of wood, B217. List of
publications on wood finishing subjects, B218. List of
publications on wood preservation, B219. List of publications relating to fungus
defects in forest products
and to decay in trees, B220.
Some reference books on
domestic and foreign woods,
B221. Wood handbook, basic
information on wood as a
material of construction with
data for its use in design
and specification, B272
────── ────── Library. Forestry
current literature, B222.
Publications on forestry,
1935-1940, B223. Subject
headings for forestry libraries, B285. A union
checklist of forestry serials,
B277
────── Laws, Statutes, etc. Farm
relief and Agricultural
adjustment acts, G30. Federal farm loan act with
amendments, and Farm
mortgage and Farm credit
acts, G31. Laws relating to
agriculture, F25. Revised
edition of laws applicable to
the United States Department
of Agriculture, 1945, F26
────── Library of Congress. Biological sciences serial publications, A119. Monthly checklist of state publications, p.
103. Scientific and technical
serial publications, United
States, 1950-1953, A120. Statistical bulletins [Census
Library Project], F20. Statistical yearbooks [Census
Library Project], F20
────── National Archives. Preliminary inventory of the records
of the Forest Service, B224.
Records of Soil Conservation
Service; preliminary check
list of records of Soil Conservation Service 1928-1942, D40
────── Office of Education. Agricultural series 17, F36. Directory of schools of agriculture
in the Latin American Republics, July 1941, F41. Notes on
research in home economics
education, E66. Summaries of
studies in agricultural education, F36. Vocational education bulletin 180, F36
────── Office of Experiment Stations.
Agricultural research institutions and library centers in
foreign countries, A79. Bibliography on poultry husbandry,
C179. Bibliography on poultry
industry, C180. Experiment
station record, A22. Federal
legislation, rulings, and regulations affecting the state
agricultural experiment stations, F27. List of bulletins
of the agricultural experiment
stations in the United States
from their establishment to
the end of 1920, A43
────── Office of Foreign Agricultural
Relations. Foreign weights
and measures with factors
for conversion to United States
units, A132

U. S. Office of Industry and Commerce. Domestic commerce series 30, B208
—— Office of Technical Services. Bibliography of reports on foods and the food industry, April 1949, E67
—— Plant Quarantine and Control Administration, history, C383
—— Production and Marketing Administration. Bibliography of marketing publications issued by the U. S. Department of Agriculture from 1937 through 1946, F13. Conversion factors and weights and measures for agricultural commodities and their products, A133. Directory of refrigerated storage warehouses in the United States, E129. List of plants approved under the poultry inspection and grading program, C203
—— Public Health Service. Division of Water Pollution Control. Public Health Service publication 270, C478
—— Quartermaster Corps. Military Planning Division. Report 129, C21
—— Quartermaster Food and Container Institute. Bibliographic series, E69. Bibliography on the deterioration of fat in foods, E68. A guide to the literature on food research and technology, G23
—— Signal Office. Bibliography of meteorology, D80
—— Soil Conservation Service. Miscellaneous publication 10, D42; 15, D44; 21, D35. Soil conservation literature, D41
  archival records, D40
—— Superintendent of Documents. Animal industry, farm animals, poultry, and dairying, C83, C181. Bibliography of United States public documents, A44. Fish and wildlife, C487. Insects: worms and insects harmful to man, animals, and plants, C318. List of publications of the Agriculture Department, 1862-1902, A44. Monthly catalog of United States government publications, p. 103. Price list 21, C487; 38, C83, C181; 41, C318
—— Surgeon General's Office. Books on veterinary medicine and related topics published during the years 1936 to 1946, C222
—— Weather Bureau. W. B. 1445, D82
United States Bureau of Fisheries, its establishment, functions, organization, resources, operations, and achievements, U. S. Bureau of Fisheries, C550
United States egg and poultry magazine, p. 84
United States food products directory, the blue book of food packers and distributors, E128
United States Record of Performance directory, C199
Ussowsky, B. N., Geminova, N. F., and Krasnosselskaja, T. C. English-Russian agricultural dictionary, A60
Utrecht. Rijksuniversiteit. Veeartsenijkundige Faculteit. Bibliotheek. Catalogus van de Veeartsenijkundige faculteit, C231

Van Dersal, W. R. Native woody plants of the United States, their erosion control and wild life values, B79
Van Royen, W. Atlas of the world's resources; v. 1, A107; v. 3, B274
Varela, A. and Blanco, G. Publicaciones periodicas agricolas en American Latina, A121
Vaughan, H. W. Breeds of live stock in America, C105
Veenman's agrarische Winkler Prins; encyclopedie vor landouw, tuinbouw en bosbouw, G11

Vegetable oil industry directory, E112
Vegetable trade directory, B118
Vegetables, B108
Verdoorn, F. Selected references on current research in plant taxonomy, ecology, and geography in Europe, Africa, Asia, and Australia, B32
Verdoorn, F. and J. G. The index botanicorum, B73
Vergani, F. and Diaz Ungria, C. Bibliografia veterinaria venezolana (fichero venezolana de bibliografia veterinaria y ganadera), C235
Vernon, A. The history and romance of the horse, C155
Veterinärmedizin, p. 91
Veterinarnyĭ entsiklopedicheskiĭ slovar', C261
Veterinary bulletin, Commonwealth Bureau of Animal Health, C218
Veterinary entomology bibliographies, C219, C236
Veterinary medicine
　abstract journals, C218-C220
　bibliographies, C221-C248
　dictionaries, C249-C263
　directories, C98, C264-C271
　encyclopedias, C243-C248
　history, C272-C287
　library catalogs, C225-C231
　military history, C282
　periodicals, C280-C289
　terminology, C249, C259
　See also Medicine
Veterinary zoology bibliography, C237
Veyret, P. Géographie de l'élevage, C163
Vicaire, G. Bibliographie gastronomique, E70
Virus diseases of plants, B166, B170-B171, B177, B182, B184-B185
Vitamin abstracts, E25
Vitamins
　abstract journals, E19, E25
　bibliographies and indexes, E34, E40-E48, E55, E60
　in foods, E150-E158

Viticulture bibliography, B92; see also Wines and wine-making
Vocational agricultural education see Agricultural education
Vogt, W. Holzfachwörterbuch, B242
Volin, L., and others. Agricultural geography of Europe and the Near East, A100
Von Loeseke, H. W. and Stephany, C. D. Selected references on yeast, G26

Wade, J. S. A bibliography of biographies of entomologists, C385. List of entomological publications of personnel of Cereal and Forage Insect Investigations, U. S. Bur. of Entomology, 1904-1928, inclusive, C316. On entomological publications of the United States government, C319. A selected bibliography of the insects of the world associated with sugar cane, their predators and parasites, C330
Wageningen, Netherlands. Bibliotheek der Landbouwhoogeschool. Bulletin-catalogus, A45
Wagner, F. H. and Hickey, J. J. A check list of technical game bulletins published by the state game departments, C22
Walker, H. J. O. Catalogue of bee books collected and offered for sale by Lt.-Col. H. J. O. Walker, C410
Wallerstein laboratories communications, E26
Wallmeyer, B. Pelztragende tiere, C40
Walsh, Benjamin Dana, bibliography, C312
Ward, A. The encyclopedia of food, E93
Wardle, R. A. Principles of applied zoology, C63
Warner, M. F., and others. A bibliography of plant genetics, B148

Washington (State). Dept. of Fisheries. Translations of fisheries literature from foreign languages into English, C497

Wasikowski, K. and Apstein, C. Periodica zoologica, C58

Water conservation
　abstract journals, D24
　bibliographies and indexes, D30, D35, G8

Weather see Meteorology

Weber, A. and Gram, E. Plant diseases in orchard, nursery, and garden crops, B183

Weber, G. A. The Bureau of Entomology, C382. Food, drug, and insecticide administration, C463. The Plant Quarantine and Control Administration, C383

Weed, C. M. A partial bibliography of the economic relations of North American birds, C27

Weeds
　bibliography, B31
　handbooks, manuals, textbooks, B77

Wehmer, C. Die pflanzenstoffe botanisch-systematisch bearbeitet, B66

Weights and measures, A128, A130-A133

Weil, B. H. and Sterne, F. Literature search on the preservation of food by freezing, E71

Weintraub, F. C. Grasses introduced into the United States, B80

Weiss, F. A. and O'Brien, M. J. Index of plant diseases in the United States, B180

Weiss, H. B. The pioneer century of American entomology, C384

Welch, P. S. Limnology, C533

Wentworth, E. N. America's sheep trails, C149

Wentworth, E. N. and Towne, C. W. Pigs, C152. Shepherd's empire, C148

Werdermann, E. and Melchior, H. rev. Syllabus der pflanzenfamilien; mit besonderer berücksichtigung der nutzpflanzen nebst einer übersicht über die florengebiete der erde, B14

West, C. J. Pulp and paper manufacture, B225

West, G. P. and Miller, W. C. Black's veterinary dictionary, C257

Westcott, C. Plant disease handbook, B188

Western canner and packer. 1952. Statistical review and yearbook number, E130

Western farm equipment directory, 1954, D75

Westwood, T. A list of books relating to fish, fishing and fisheries to supplement the Bibliotheca piscatoria ... , C494. Bibliotheca piscatoria, C493

Wetzel, C. M. American fishing books, C495

Whaling
　bibliography, C480
　history, C538, C540, C548

Whalley, M. E. Bibliography on the influence of mineral deficiencies on growth and reproduction of farm animals, C75

What do studies show? Summaries and interpretations of research in selected areas of agricultural education, F43

Wheat, B144

Wheeler, L. B. and Amerine, M. A. A check list of books and pamphlets on grapes and wine and related subjects, 1938-1948, E28

Wheeler, W. A. Forage and pasture crops, B126

Whipple, C. E., and others. Agricultural geography of Europe and the Near East, A100

White, J. A compendious dictionary of the veterinary art, C262

White, J. M. The farmer's handbook, A82

White cattle, C123, C125

Whitehead, G. K. The ancient white

Whitehead, G. K. (continued) cattle of Britain and their descendants, C125

Whiteley, M. A. and Thorpe, J. F. Thorpe's dictionary of applied chemistry, D17

Whitlock, C. Abbreviations used in the Department of Agriculture for titles and publications, A108

Whitney cookery collection, The. New York Public Library, E53

Who's who in America, p. 110

Who's who in meat, a directory of the meat industry, E131

Who's who in the dairy industries, 1953, E132

Who's who in the egg and poultry industries, C205, E133

Wiebe, R. and Nowakowska, J. The technical literature of agricultural motor fuels including physical and chemical properties, engine performance, economics, patents, and books, D63

Wieland, L. H. Bibliography on soil conservation, D42. Soil erosion bibliography, D43

Wiesner, J. von. Die rohstoffe des pflanzenreichs, B67

Wilcox, E. V. Modern farmers' encyclopedia of agriculture, A51

Wildlife
  abstract journals, C4-C5
  bibliography, C26
  conservation, C47, C49, D35, D49
  directories, C47, C49, D49

Wildlife abstracts, 1935-51, U. S. Fish and Wildlife Service, C5

Wildlife refuge bibliography, C20

Wildlife review, U. S. Fish and Wildlife Service, C4

Willham, O. S. A genetic history of the Hereford breed of cattle in the United States, C138

Williams, A. C. Dictionary of trout flies and of flies for sea-trout and grayling, C513

Willis, J. C. Dictionary of the flowering plants and ferns, B48

Wilson, H. F. The historical development of insect classification, C375

Wilson, J. The evolution of British cattle and the fashioning of breeds, C126

Winchell, C. M. Guide to reference books, A8

Wine Institute, San Francisco. Selective bibliography of wine books, E72

Wines and vines. Annual directory issue, 1954, E134

Wines and wine-making
  abstract journals, E24
  bibliographies and indexes, E28, E57-E59, E70, E72
  dictionaries, E74-E75, E86, E87a, E89-E91
  directories, E99, E106, E110, E121-E122, E134, E146
  handbooks, manuals, textbooks, E146

Winton, A. L. and K. B. The structure and composition of foods, E148

Wirth, D. Lexikon der praktischen therapie und prophylaxie für thierärzte, C263

Wodehouse, R. P. Hay fever plants, B81

Wisconsin. Department of Agriculture. Wisconsin registered licensed veterinarians, C271

—— University. College of Agriculture. Library. A list of the publications on apiculture contained in the Dr. Charles C. Miller Memorial Apiculture Library, C411

Wolf, F. A. and F. T. The fungi, B198

Wolfe, L. S. Farm glossary, A61

Wood
  bibliographies and indexes, B213-B214, B216-B221
  chemistry, B212
  dictionaries, B229-B230, B233, B237
  See also Forestry. Dictionaries

Wood (continued)
   handbooks, manuals, texts,
      B267, B269, B272
   preservation, B217, B219
   statistics, B278-B282
Wood-decaying fungi, B220
Wood finishing, B218
Wood-pulp, B215, B225, B226
Woods, R. S. Addenda to the
   Naturalist's lexicon, C41.
   The naturalist's lexicon,
   C42
Woods, R. S. and Marshall, W. T.
   Glossary of succulent plant
   terms, B44
Wool bibliography, C72, C76
Wooster, H. A. Nutritional data,
   E154
Workers in subjects pertaining to
   agriculture in land-grant
   colleges and experiment
   stations, 1954-1955, U. S.
   Department of Agriculture,
   A78
World bibliography of bibliographies and of bibliographical catalogues, calendars, abstracts, digests, indexes, and the like, T. Besterman, A1
World fisheries abstracts, C468
World Health Organization. Insecticides, manual of specifications for insecticides and for spraying and dusting apparatus, C460. World medical periodicals, C289
World list of bee research workers, C427
World list of scientific periodicals published in the years 1900-1950, A122
World medical periodicals, C289
World's poultry science journal, p. 84
Worms see Helminthology
Wright, R. P. Standard cyclopedia of modern agriculture and rural economy, A52
Wyman, D. The arboretums and botanical gardens of North America, B51

Yale University School of Forestry. Bulletin 40, B234
Yates, F. and Fisher, R. A. Statistical tables for biological, agricultural, and medical research, A127
Yeager, L. E. A contribution toward a bibliography on North American fur animals, C28
Yearbook of agriculture, U. S. Department of Agriculture, A81
Yearbook of fishery statistics, C556
Yearbook of food and agricultural statistics, Food and Agriculture Organization of the United Nations, F15
Yeast, G25, G26
Youatt, W. Cattle: their breeds, management, and diseases, C127. A history of the horse, C156. Sheep, C150

Zander, R. and Heckel, M. Wörterbuch der gärtnerischen fachausdrücke in vier sprachen, B110
Zeitschrift für fischerei und deren hilfswissenschaften, C470
Zeitschrift für lebensmittel-untersuchung und -forschung, E27
Zeitschrift für pflanzenernährung, düngung und bodenkunde, D25
Zeitschrift für pflanzenkrankheiten (pflanzenpathologie) und pflanzenschutz, B163
Zeitschrift für pflanzenzüchtung, B131
Zentralblatt für bakteriologie, parasitenkunde und infektionskrankheiten, B164
Ziegler, H. E. Zoologisches wörterbuch, C43
Zimmerman, F. L. and Read, P. R. Numerical list of current publications of the United States Department of Agriculture, A46
Zoogeography see Zoology, Economic. Atlases and geography
Zoological record, C10
Zoological Society of London. List of the periodicals and serials

Zoological Society of
London (continued)
   in the Library ... , C61
Zoologischer anzeiger, p. 101
Zoologisches adressbuch, C50
Zoology, Economic
   abstract journals, A11, B130, C3-C5
   atlases and geography, C51-C56
   bibliographies of bibliographies, C1-C2
   bibliographies and indexes, A2, A24, C6-C26, C64, C237
   dictionaries, C29-C43
   directories, C47-C50
   handbooks, manuals, textbooks, C2, C62-C63
   nomenclature, C1, C10, C29, C38, C44-C46
   periodicals, C57-C61
Der züchter; zeitschrift für theoretische und angewandte genetik ... , B132

www.ingramcontent.com/pod-product-compliance
Lightning Source LLC
Chambersburg PA
CBHW021704230426

43668CB00008B/714